The Fertile Sea

The Fertile Sea

being The Buckland Lectures for 1957

by A. P. Orr, M.A., D.Sc., F.R.S.E.
and S. M. Marshall, D.Sc., F.R.S., F.R.S.E.
of
The Marine Station, Millport

Fishing News (Books) Limited
110 Fleet Street, London, E.C.4

A PIONEER
OF FISHERY RESEARCH

Frank Buckland 1826–1880

This is a Buckland Foundation book—one of a series providing a permanent record of annual lectures maintained by a bequest of the late Frank Buckland

First published 1969

Printed by The Whitefriars Press Limited, London and Tonbridge

Contents

List of Illustrations

List of Plates (*between pp. 64 & 65*)

Acknowledgements

Dr A. P. Orr delivered the Buckland lectures for 1957 but he died before he had prepared them for publication. He had asked me to collaborate with him in this, we had often discussed how to do it and I have now tried to complete the preparation on those lines.

I am grateful to Sir Frederick S. Russell for reading and helpfully criticizing the typescript.

I am also grateful to the many authors and publishers who have allowed me to use their illustrations.

Dr C. M. Yonge, Dr D. P. Wilson, Dr J. H. Fraser, Dr H. G. Vevers, Dr C. F. Hickling and Dr M. R. Droop have kindly lent negatives or photographs. To the following I am indebted for permission to copy illustrations: the Council of the Marine Biological Association, U.K. (for many text-figures from their Journal and from Dr M. V. Lebour's book *Dinoflagellates of Northern Seas*); Messrs Collins (for illustrations from *The Sea Shore* and *Oysters*, by Dr C. M. Yonge, and *The Open Sea*, by Sir Alister Hardy); Messrs. Oliver and Boyd (for illustrations from *The biology of a marine copepod* Calanus finmarchicus); The Cambridge University Press (for Fig. 1, from *The Chemistry and Fertility of Seawater*, by Dr H. W. Harvey); the Indo-Pacific Fisheries Council (for Plate XII (lower)); the Bingham Oceanographic Laboratory (for Fig. 33); the Controllers of H.M. Stationery Office (for Figs. 17 and 34); The Pergamon Press Ltd. (for Fig. 11, from *Plankton and Productivity in the Oceans* 1963, by Dr J. E. G. Raymont); and Messrs G. T. Foulis (for Fig. 18, from *Nature Adrift*, by Dr James Fraser).

I am glad to acknowledge the help of Miss Heather Warwick who made many of the drawings and that of Mr A. Elliott with the photography.

It is impossible to give references for all the work mentioned and the books and journals used, but some books which give a comprehensive treatment and a few papers of fundamental importance are listed at the end.

Chapter 1

The Sea as a Medium for Plant Growth

> "There," said he, pointing to the sea, "is a green pasture where your children's grandchildren will go for bread".
>
> OBED MACY, *History of Nantucket*

The sea covers almost three quarters of the earth's surface and contains in its upper illuminated layers myriads of microscopic plants, the phytoplankton, on which all the animals in the sea, including the fishes, depend directly or indirectly for their food. It has been calculated that in spite of their small size these plants are so numerous that in bulk their production over the oceans exceeds that of the plants on the land. Here we shall describe the conditions affecting plant growth in the sea. Among the most important of these are nutrient salts, light, temperature and circulation.

Sea water contains in solution about 3·5% of salts; the concentration varies slightly with the latitude but is rarely less than 3·3 or more than 3·8% except in inshore or partly enclosed areas. The chief ions present are those of sodium, magnesium, calcium, potassium, chloride, sulphate, bromide and carbonate, and together these account for more than 99·9% of the salts present. The ratio of these ions to one another shows a remarkable constancy, irrespective of the total percentage of salts in solution. It is customary to express the salt content as the salinity, which in 1901 was defined as the total weight in grammes of solid matter dissolved in a kilogram of sea water. This is an estimation very difficult to make accurately and it was usual to derive the salinity from a titration of the chlorides with silver nitrate. Recently physical methods have been brought into use. Salinity can now be estimated from a measurement of the electrical conductivity of the sea-water sample, from its density, its refractive index, or the velocity of sound through it. In Table 1 are given the major constituents of sea water of salinity 34·5‰ (i.e., parts per thousand).

TABLE 1

The major constituents of sea water with a salinity of 34·5‰

Ion	Grammes per kilo	%
Na^+	10·556	30·61
Mg^{++}	1·272	3·69
Ca^{++}	0·400	1·16
K^+	0·380	1·10
Sr^{++}	0·013	0·04
Cl^-	18·980	55·04
SO_4^-	2·649	7·68
HCO_3^-	0·140	0·41
Br^-	0·065	0·19
H_3BO_3	0·026	0·07
Total	34·481	99·99

Apart from the major constituents, almost every element has been shown to be present in sea water and the concentrations of those known are given in Table 2 where the values are for sea water from depths less than 1000 m. All of the major constituents and an ever-increasing number of the minor ones have been found to play a role in plant production in the sea. In addition to the major and minor constituents, sea water also contains organic substances in solution. These are still very little known, but both from direct and indirect evidence some have been shown to be essential for plant growth.

The influence of the major constituents of sea water on plant life has, apart from carbon dioxide, been relatively little studied. More is known about their effect on inshore plants, chiefly flagellates, but there is information also on the marine diatom *Skeletonema costatum* and on the oceanic *Rhodomonas lens*. Contrary to expectation, changes in the salt content are not

TABLE 2

Minor elements in sea water

Element		Mg/l	Element		Mg/l
Aluminium	Al	0·01	Mercury	Hg	0·00003
Antimony	Sb	0·0005	Molybdenum	Mo	0·01
Arsenic	As	0·003	Neon	Ne	0·0003
Barium	Ba	0·0062	Nickel	Ni	0·0005
Bismuth	Bi	0·0002	Nitrogen	N	0·5
Cadmium	Cd	0·04	Phosphorus	P	0·07
Caesium	Cs	0·0005	Radium	Ra	$3{\cdot}0 \times 10^{-11}$
Cerium	Ce	0·0004	Radon	Rn	$9{\cdot}0 \times 10^{-15}$
Chromium	Cr	0·002	Rubidium	Rb	0·12
Cobalt	Co	0·0005	Scandium	Sc	0·00004
Copper	Cu	0·003	Selenium	Se	0·004
Fluorine	F	1·3	Silicon	Si	3
Gallium	Ga	0·0005	Silver	Ag	0·0003
Geranium	Ge	0·0001	Thallium	Tl	0·00001
Gold	Au	0·000004	Thorium	Th	0·0007
Helium	He	0·00005	Tin	Sn	0·003
Indium	In	0·02	Titanium	Ti	0·501
Iodine	I	0·05	Tungsten	W	0·0001
Iron	Fe	0·01	Uranium	U	0·003
Lanthanum	La	0·0003	Vanadium	V	0·002
Lead	Pb	0·003	Yttrium	Y	0·0003
Lithium	Li	0·2	Zinc	Zn	0·01
Manganese	Mn	0·002			

critical and the optimum concentrations are not closely related to values in the sea.

It has been found that divalent sulphur or organic sulphur compounds are necessary for diatom growth, and in some freshwater lakes in Africa sulphur appears to be insufficient.

Apart from traces of organic compounds in solution, all the carbon in sea water is present as carbon dioxide, bicarbonate and carbonate. The interrelationships of these depend on the temperature, the hydrogen ion concentration and the excess base (Fig. 1). Free CO_2 is maximal at pH 4 when the water is on the acid side, and decreases as it becomes more alkaline; it is almost zero at pH 9. The reverse is true of carbonate.

Natural sea water is normally slightly alkaline and the pH value usually varies within rather narrow limits, from 7·9 to 8·3

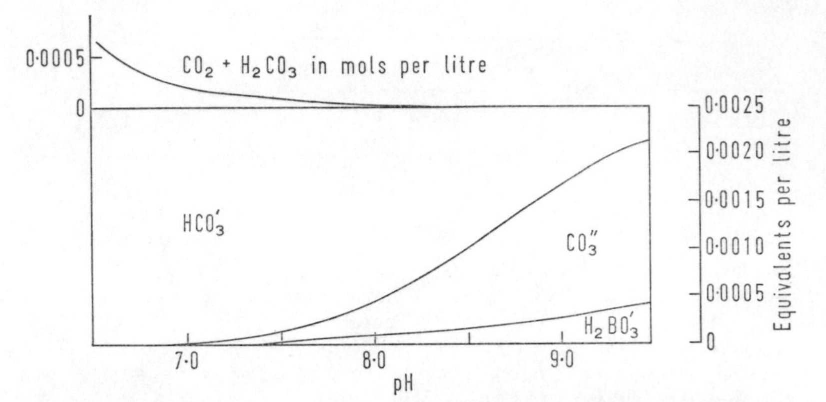

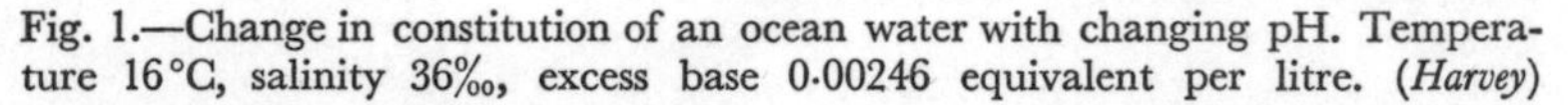
Fig. 1.—Change in constitution of an ocean water with changing pH. Temperature 16 °C, salinity 36‰, excess base 0·00246 equivalent per litre. (*Harvey*)

in open sea conditions. As carbon dioxide is removed from the water by the plants, the equilibrium is altered and the pH value increases. The annual cycle in plant growth is reflected in the changes in pH value, and for a station in the English Channel these are shown in Fig. 2. From the changes in the carbon dioxide equilibria it is possible to calculate approximately the production of phytoplankton.

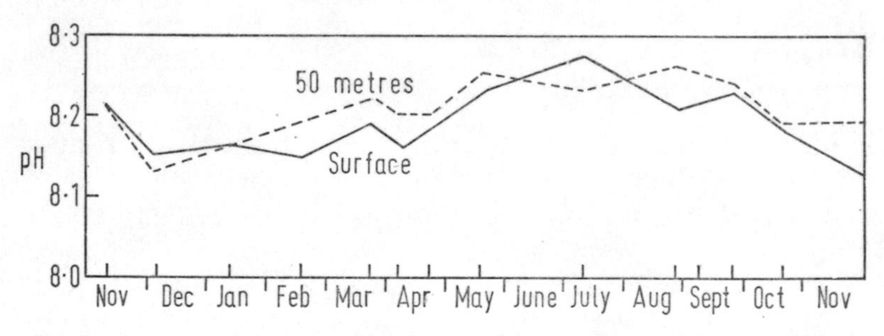

Fig. 2.—Seasonal changes in pH at Station L4 in the English Channel, 1930–31.

The most important minor constituents for the phytoplankton are phosphorus and nitrogen compounds and to a less degree silicon. The unexpected importance of this last element is because the skeleton of the diatoms, the most important members of the phytoplankton in temperate and polar waters, is siliceous.

The necessity of these minor constituents for plant growth in the sea was shown by workers of the German school at Kiel. Their methods were however cumbersome and progress was slow till workers at the Plymouth laboratory developed rapid and accurate colorimetric methods.

The amount of phosphorus in solution in sea water is small and varies with the latitude, the depth and the season. In the sea it normally occurs as orthophosphate but it has been shown that there is always a certain proportion present as dissolved organic phosphorus. This proportion is higher during the spring and summer and lower in the winter. There may indeed be as much dissolved organic as inorganic phosphorus in early summer in the English Channel.

Observations on the annual cycle in the dissolved inorganic phosphorus have been made in several places but the most complete data are those from the English Channel. The data for several years are shown in Figs. 3 and 5. In general the winter maximum in the English Channel lies between 0·32 and 0·69 μg atom P/l, but in summer it falls to 0·016–0·10 μg atom P/l.

The seasonal variation in temperate waters is linked with vertical mixing and the utilization by the phytoplankton. During the winter the surface waters are cooled by the air and thus become denser and sink, setting up a vertical circulation which makes the water column isothermal. The sea is less stable than in summer and the mixing is accelerated by strong winds. In spring the increasing light leads to a blooming of the phytoplankton and the increase in temperature promotes the formation of stable conditions, with the less dense warmer water near the surface. The phytoplankton depletes the phosphate, at first near to the surface and then gradually deeper down as the light intensity and duration increase during the late spring and summer. The depletion of the nutrients in the upper layers gradually slows down and eventually stops phytoplankton growth, though disturbances of the stable layering by wind or currents may cause a reintroduction of dissolved phosphorus into the illuminated layers and further phytoplankton production. In the autumn the surface water cools and vertical mixing begins again. Phosphate-rich deep water is thus brought into the illuminated layers. A second blooming of the phytoplankton

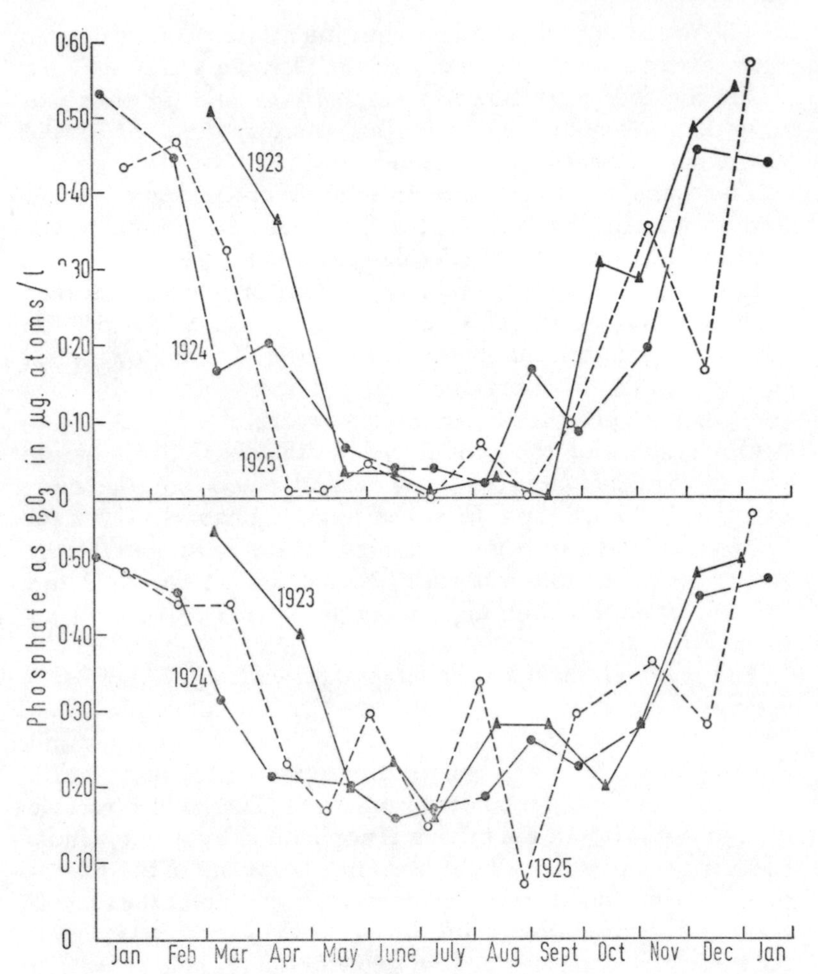

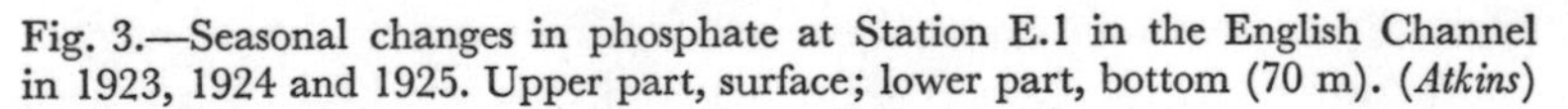

Fig. 3.—Seasonal changes in phosphate at Station E.1 in the English Channel in 1923, 1924 and 1925. Upper part, surface; lower part, bottom (70 m). (*Atkins*)

may then take place though it is usually less marked than the spring one.

In the surface layers coastal waters are generally richer in dissolved phosphate than offshore waters. In tropical and temperate seas the top 50 metres are usually low in phosphate and this poverty may extend as deep as 100 m or more. Under some conditions however, e.g. upwelling of deep water, the

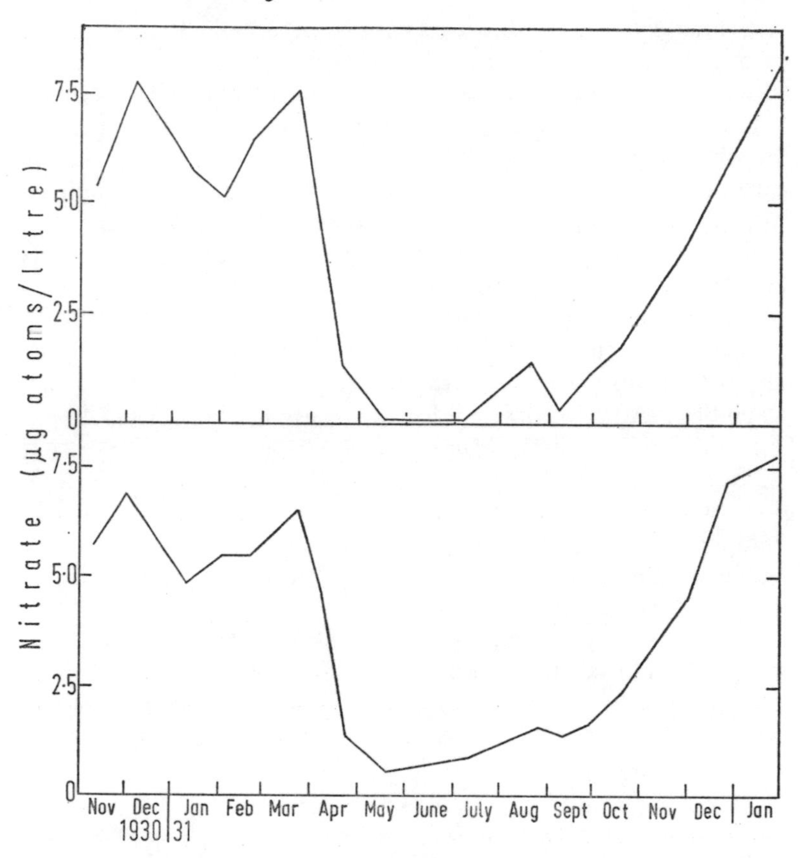

Fig. 4.—Seasonal changes in nitrate (including nitrite) at Station E.1 in the English Channel in 1931. Upper part, surface; lower part, bottom (70 m). (*Cooper*)

surface water may have enough dissolved phosphate to cause rich plant production even in low latitudes.

The Arctic and Antarctic oceans usually have a high phosphate content in the surface layers (up to 1·93–2·26 μg atoms P/l or more). Even the rich phytoplankton growth in the polar summer does not remove it completely and light seems to be the limiting factor for plant growth. Stability in summer is less marked there than in temperate or tropical waters and this too will affect the nutrient content.

Most of the combined nitrogen in the sea is present as nitrate. Nitrite and ammonia are present as well but in small amounts relative to the nitrate. The seasonal changes in the nitrogen compounds near the surface follow a course fairly similar to that of phosphorus; the annual cycle in the English Channel is shown in Fig. 4. There is a steep fall with the rise of phytoplankton production and stabilization of the sea in spring; during the summer nitrate is practically absent from the surface layers and is greatly reduced as deep as 30 m.

Decay of the plants gives rise to small quantities of ammonia just below the photosynthetic zone and this is rapidly oxidized to nitrite and then to nitrate.

By weight there is usually 7–8 times as much nitrogen as phosphorus in sea water, although this ratio is subject to much variation. Data for nitrate, along with nitrite, taken at different latitudes show that its distribution is closely parallel with that of phosphate, for it is rich at all depths at high latitudes and low or absent in the surface layers (down to 100–200 m) in temperate and tropical latitudes. In the depths of the ocean, or in the surface waters of the Arctic and Antarctic, there may be 14·3–39·3 μg atoms N/l.

The estimation of nitrogen compounds in sea water is not so simple as that of phosphorus and we have fewer seasonal data. After a phytoplankton outburst the nitrate is often completely exhausted and it therefore tends to act as a limiting factor more often than phosphate.

Organic forms of nitrogen are also present but the analyses show variable results and the seasonal changes are unknown.

Finally it is possible that some nitrogen may be obtained by the fixation of atmospheric nitrogen. Blue-green algae can do this and, in their excretions, make more nitrogen available for the rest of the phytoplankton.

Silica, which is an important constituent of the tests of diatoms, also shows a seasonal variation, as would be expected. In the surface waters, during the summer, values as low as 0·35 μg atom Si/l have been recorded, while in deep water values of over 35·3 μg atoms/l are general. The annual cycle in the English Channel is shown in Fig. 5; in winter, values are about 7·1 μg atoms/l and in summer near the surface they fall to less than 1·1 μg atoms/l. From a study of the silicate figures for the English Channel, it has been suggested that the silica in diatoms may be rapidly liberated on their death and may be utilized several times by successive generations in the course of a single season. Against this is the fact that large areas of the ocean bottom are covered with diatom ooze in which the individual diatom skeletons are apparently unaltered.

Although a very large number of other elements have been detected in sea water, little research has so far been done on the requirements of these by plants or on their seasonal variations. It

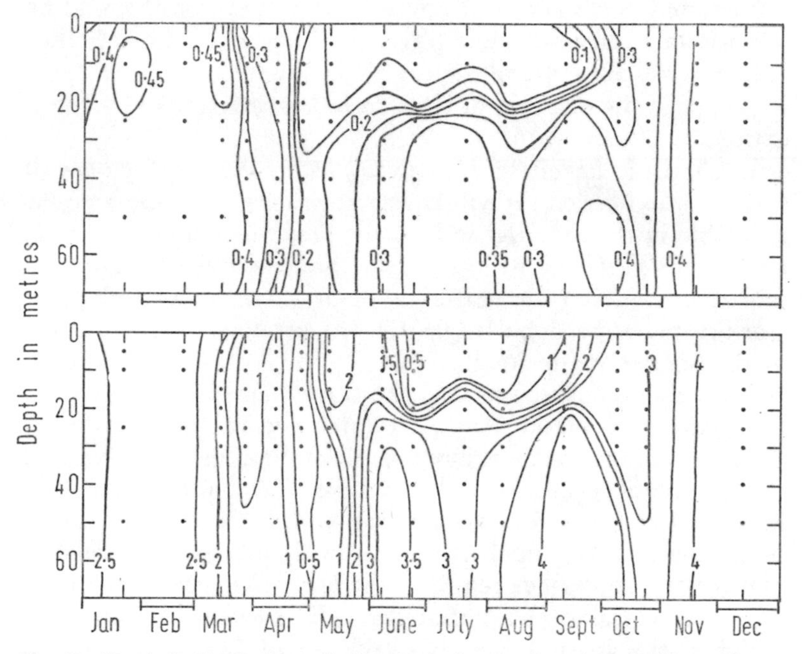

Fig. 5.—Vertical distribution of phosphate (upper figure) and silicate (lower figure) as μg atoms/l at Station E.1 in the English Channel in 1953. Contour lines at 0·05 μg atom intervals. (*After Armstrong*)

is known that traces of zinc, manganese, copper and cobalt are required in phytoplankton cultures, that gallium stimulates cell production when nitrogen and phosphorus compounds are deficient, and that molybdenum is necessary for some blue-green algae.

We know more of the iron requirements of phytoplankton. In sea water the solubility of this element is very low, less than 10^{-7} mg/l, but diatoms can probably utilize colloidal iron. Manganese, in concentrations of 1 part in 10^9 parts of sea water, causes vigorous growth of the marine diatom *Ditylum brightwellii*. Iodine is present in the littoral algae and part of it is bound in the organic form.

Vitamin B12 is essential for the growth of at least half of the marine flagellates and diatoms that have been tested. The amount present in the sea is very small, varying from $0{\cdot}2\text{–}5{\cdot}0 \times 10^{-6}$ mg/ml, and since the lowest values are obtained in the illuminated upper layers, it might be expected that it would be of importance in controlling phytoplankton production. Higher values in the oxygen-poor layers below the photosynthetic zone suggest that the presence of vitamin B12 is linked with bacterial action.

In the little flagellate *Monochrysis lutheri* the requirement of this vitamin for rich growth is very much less than the amount available in the sea, and indeed the facts suggest that there is always more vitamin B12 in the sea than is required. However, more is present in inshore than in open waters, and much more work needs to be done before the importance of vitamin B12 in the sea can be decided.

The minor constituents and the trace elements necessary for the growth of plants are sometimes present in such small quantities in the sea that they may limit plant growth. Even so, many instances are recorded where plant production has decreased or stopped in spite of the known minor constituents being available in sufficient quantities. It has been shown that plants have an *absolute* requirement for each nutrient, without which they cannot live; that there is a *normal* requirement, which is the amount the cells take up during active growth when none of the nutrients is limiting; and also that there is a *minimum* requirement, which is the quantity of the nutrient in a cell when that nutrient is limiting the growth of the population.

Phosphorus-deficient cells can be produced by permitting cultures to grow after removal of all the phosphate from the medium, and if these cells are then supplied with phosphate they can take it up in the dark though normal cells do not. At the other extreme, in cultures with excess phosphate, the cells store it. Under natural conditions the cells produced in the early stages of a phytoplankton increase (before phosphate is lowered by plant growth) have a higher phosphate content than those produced later.

Probably the behaviour of cells with respect to nitrate is similar to that with phosphate. Indeed it can be shown that when the nitrate-N falls below 0·0014 μg atom/l, the rate of uptake by the diatom *Nitzschia closterium* is directly related to the concentration.

The ratio in which the nutrients are removed from sea water is C:N:P as 100:16·7:2·5. For N and P this approximates to the ratio in which these are present in sea water and in the phytoplankton itself. The ratios "are comparable to, though much less precise than, the combining proportions of chemistry".

The use of cultures of marine algae has proved a very fruitful method for studying experimentally the conditions for plant growth in the sea. Allen and Nelson, who were pioneers in this subject, showed that "persistent" cultures of many species of diatoms could be maintained in the laboratory in artificial media. These cultures had only one species of diatom but contained bacteria and a few flagellates. Allen and Nelson found, however, that it was essential to add a small quantity of natural sea water to get successful growth. This was because most diatoms require "accessory growth factors" which occur in natural sea water.

The media first used were found to be improved by adding to the enriched sea water sterile soil extract. Using this "Erd-Schreiber" medium a much larger variety of algae has been successfully cultured by many different workers, and it is probably the medium used most often for general work at present. The action of the soil extract may be due in part to the trace metals present, or to organic growth factors, or even to its buffering capacity. Its defects are that it is not readily reproducible since the soils used may be widely different.

A more recent development of culture work has been an attempt to find the precise nutrient requirements of different species of algae. Trace metals have been investigated by several workers. A non-toxic reservoir of these can be supplied by adding a suitable quantity of a chelating agent, such as ethylene diamine tetra-acetic acid, so that the amounts of each available in solution can be controlled.

A number of organic substances have also been found necessary and again certain algae have their own specific requirements. In bacteria-free cultures *Skeletonema costatum* requires vitamin B12. Organic compounds containing divalent sulphur (e.g. thiamine) are necessary for continuous growth of *Ditylum brightwellii*, and cysteine is stimulating for *Skeletonema*. There is a rapidly expanding literature on this subject.

Photosynthesis by the phytoplankton depends on the amount of light it receives, which in turn depends on the latitude, the time of year, the depth in the water and the turbidity. The variation in light energy with latitude throughout the year is shown in Fig. 6. The amount of radiation which reaches the

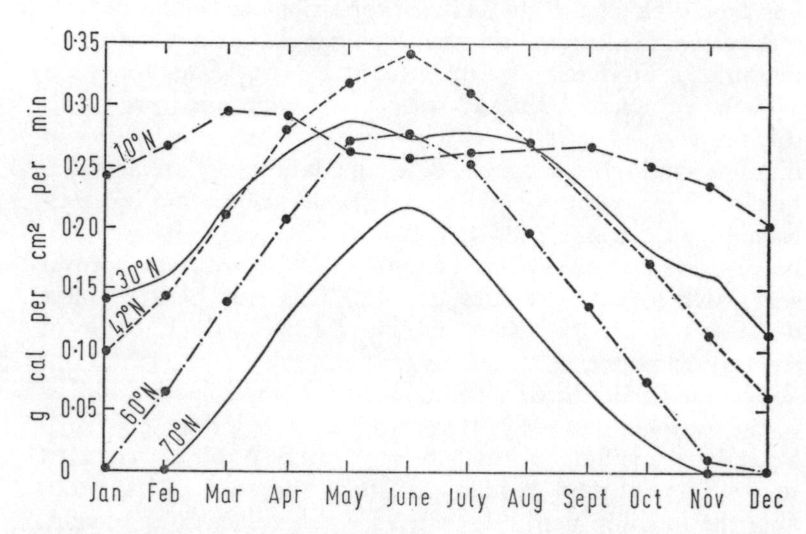

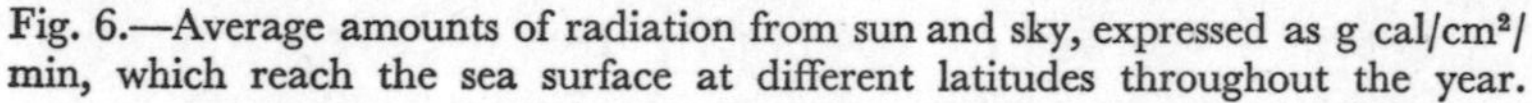

Fig. 6.—Average amounts of radiation from sun and sky, expressed as g cal/cm²/min, which reach the sea surface at different latitudes throughout the year.

sea surface depends also on the angle of the sun, and on absorption and reflection in the atmosphere by clouds and by other scattering. Cloudiness is probably the most important factor for absorption and reflection. On cloudless days about 80% of the incident radiation may reach the earth's surface, but with heavily overcast conditions it may be 20% or even less; the average figure is about 65%. The total radiation received over the year at different latitudes has been calculated and it turns out that between April and August the difference with latitude from the equator as far N as 60° is small.

Chapter 2

Phytoplankton

Diatoms have lacy cases
Of material siliceous,
Perforated lids and bases
Made to fit like Petri dishes.
Nursed on Nature's hydroponic
They're prolific and nutritious,
Making bouillabaisse planktonic
For the sustenance of fishes.

RALPH LEWIN

On land, plants derive their nourishment from the air and from the soil in which they are rooted. Marine algae have no roots and are immersed most or all of the time, so they must get their nutrients from the substances dissolved in the sea water. Both on land and in the sea the light of the sun is essential to plants for photosynthesis.

The most important of marine plants are not, as one might expect, the seaweeds, which form only a narrow fringe round the coasts, but minute one-celled organisms known collectively as phytoplankton. Most of them are invisible to the naked eye as individuals, although when present in enormous numbers, as they often are, they colour the water brown or red or green.

Some of these unicellular plants belong to the same algal divisions as the larger seaweeds. The green algae (Chlorophyceae) and the blue-green (Cyanophyceae) both have their planktonic representatives; numerous pelagic forms belong to the yellow-green algae or Chrysophyceae, and the most important group, at least in cold and temperate waters, is that of the diatoms (Bacillariophyceae).

A diatom is enclosed in a shell of silica, divided in two pieces fitting into one another rather like a pillbox and its lid

(Fig. 7). Most diatoms are pelagic, that is they spend their whole existence floating in the water, but some live attached to the bottom, or to the surface of sedentary or floating objects. It is the first kind which play the biggest part in the productivity of the sea and, because they have to stay up in the well-lit surface layers, their siliceous shells are much lighter than those of bottom-living forms. These shells or frustules have long been used as objects to test the resolving power of microscope lenses, because of the elaborate pattern of lines and dots which they show. The electron microscope now shows that the frustules of planktonic diatoms are really composed of two layers, each being a rather open network of siliceous threads, where holes make up as much as 50% of the area. The attached (or sessile) forms have heavier, one-layered shells with only about 10% of the area as holes.

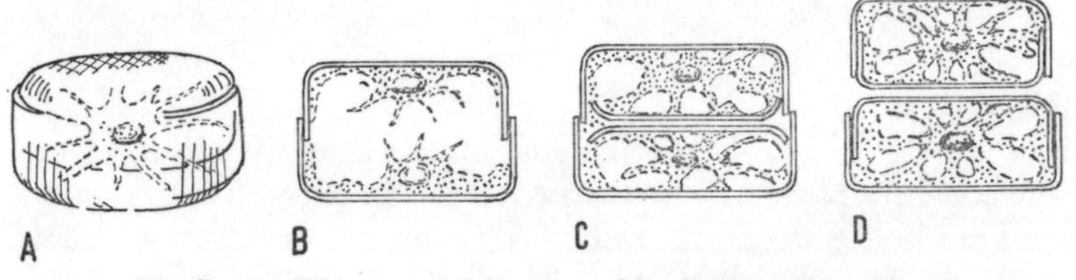

Fig. 7.—A pillbox type of diatom and its division. (*From Hardy*)

The difference is of course related to the fact that pelagic diatoms have to stay floating near the surface, because the strength of light diminishes quickly as it penetrates deeper into the water, and below 50–100 metres (in our latitude) there is not enough for photosynthesis. A thinner shell is not the only adaptation a diatom has to make floating easier. Its very small size helps, for the smaller an object is, the more surface it has in relation to its volume, and the greater will be the frictional resistance to sinking. A relatively large surface also makes for easier absorption of light and nutrients. In addition to its small size, a diatom shell is often provided with spines or threads (thus increasing the surface still more) and the single cells often become attached to one another in long twisting chains (Plate I). Furthermore diatoms can secrete a mucus or slime which comes out through the pores in the frustules in long threads and also helps to keep them suspended in the water.

Other groups of phytoplankton organisms do not have so much difficulty in remaining suspended, for they are provided with one or more whip-like threads or flagella which, beating to and fro or with a rotary motion, enable them to move about. Most of them are very small and delicate; on preservation they break up or alter shape and become unrecognizable. This is why less is known about their abundance and distribution in

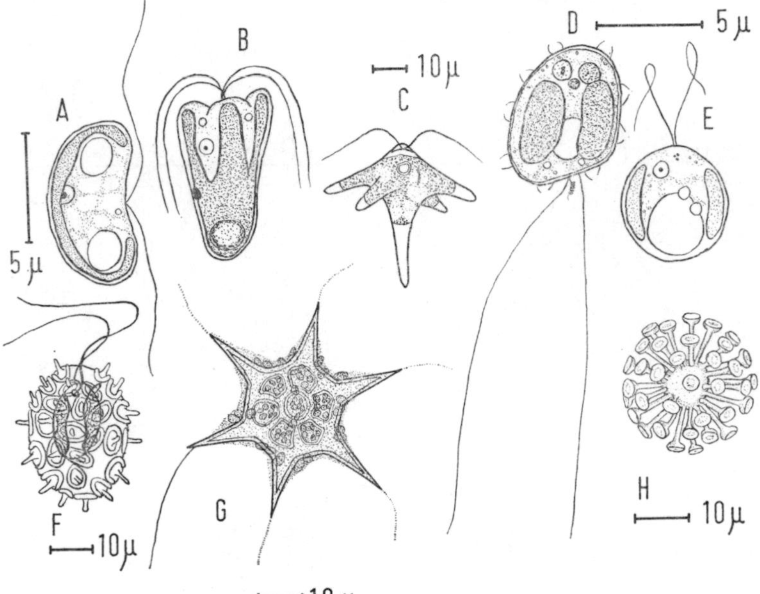

Fig. 8.—Planktonic flagellates. A, *Hemiselmis rufescens*; B, *Pyramimonas grossii*; C, *Brachiomonas submarina*; D, *Chrysochromulina chiton*; E, *Dicrateria inornata*; F, *Syracosphaera subsalsa*; G, *Distephanus speculum*; H, *Discosphaera thomsoni*.

the sea, for they are too small to be retained in tow-nets and until recently they were little studied in water samples. Now we realize that they are very useful as food for many larval forms, particularly molluscs, and a more determined effort is being made to describe and classify them. This is an essential preliminary to finding out what part they play in the productivity of the sea (Fig. 8).

Several groups of these algae belong to the Chrysophyceae. Some have, besides two flagella, a thread-like retractile organ with an adhesive tip by which they can attach themselves temporarily to solid objects (Fig. 8(D)). They are covered with very delicate ribbed scales, some with central spines, and they feed both as plants and animals. Like diatoms, these cells can exude mucous threads.

Another Chrysophycean family, the Coccolithophoridae, is

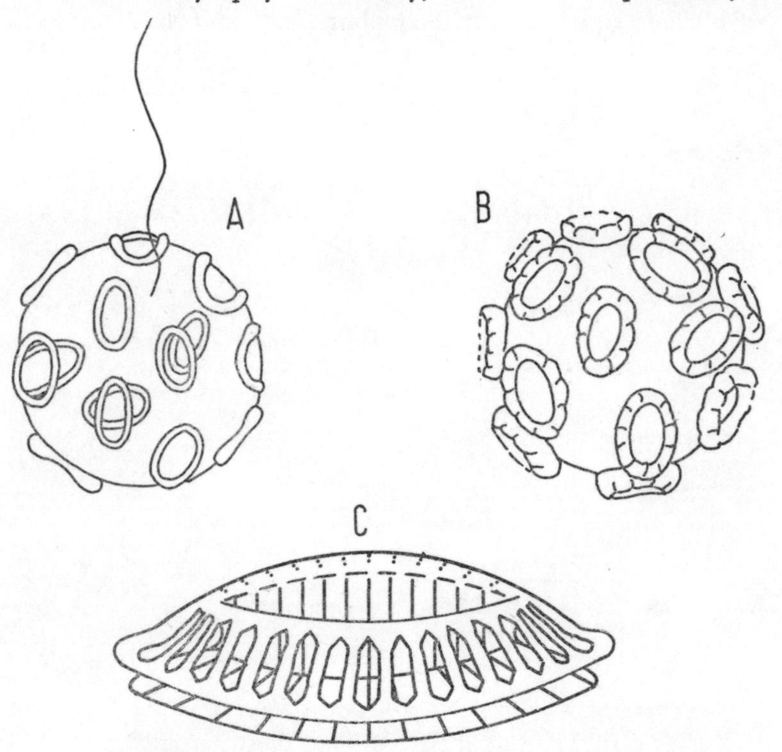

Fig. 9.—*Coccolithus huxleyi* as drawn at different times. A, 1902 (*after Lohmamm*); B, 1941 (*after Kamptner*); C, 1952, model of a single coccolith based on electron microscopy (*after Braarud et al.*).

more common in warm and oceanic waters. Each cell has two flagella and is protected by a covering of calcareous plates which take all sorts of fantastic shapes (Fig. 8(F,H)). It was not until these plates had been examined with an electron microscope that we understood how intricate was their design. Even the simplest of them, which under the highest power of the ordinary microscope looks like an open ring, is found to be made up of dozens of different pieces, elaborately sculptured and fitting together rather like a deep-rimmed platter (Fig. 9). Each single ring may be only $\frac{1}{2}\mu$ wide (1μ is equal to one thousandth of a millimetre). Nobody can as yet tell the reason for the complication of these minuscule structures, but the plates themselves are said to reflect the sun's rays and prevent damage to the cell. Certainly when coccolithophores are present in enormous numbers, as they sometimes are even in our seas, they reflect the light and make the water appear a dazzling green.

Probably belonging to the same division of algae are the Silicoflagellates *Distephanus* (Fig. 8(G)) and *Dictyocha*. These are among the prettiest of the minute flagellates. Each cell is not more than 25–30μ in diameter and has a delicate basket-work skeleton of silica of distinctive shape; which is often found double, two mirror-image skeletons fitting closely to one another by their basal rings; this is the prelude to division. Far more silicoflagellates are known as fossil skeletons than as living plants and, except sometimes in Arctic seas, they are never very common, but they are a food of some of the smallest of zooplankton groups, the Tintinnids.

The dinoflagellates (Fig. 10) belong to another great algal division, the Dinophyceae. On the whole they are larger than flagellates belonging to the Chrysophyceae. Of their two flagella, one works in a groove running horizontally round the body and the other in a vertical groove. They form a large group but by no means all of them behave as normal plants. Some are naked, some have an outer casing formed not of silica but of a number of horny plates. Some have chloroplasts and depend on photosynthesis. Some absorb dissolved nutriment from the water and some engulf and digest plants or animals smaller than themselves. It is only those which photosynthesize that can properly be called phytoplankton.

Although dinoflagellates have flagella and so can move themselves about, at least for a small distance, yet we do find mechanisms similar to those of diatoms. This is especially noticeable if related species from cold and warm water are compared. Warm water is lighter and less viscous than cold and therefore more difficult to float in; some tropical species of dinoflagellate have developed long horns and parachute-like

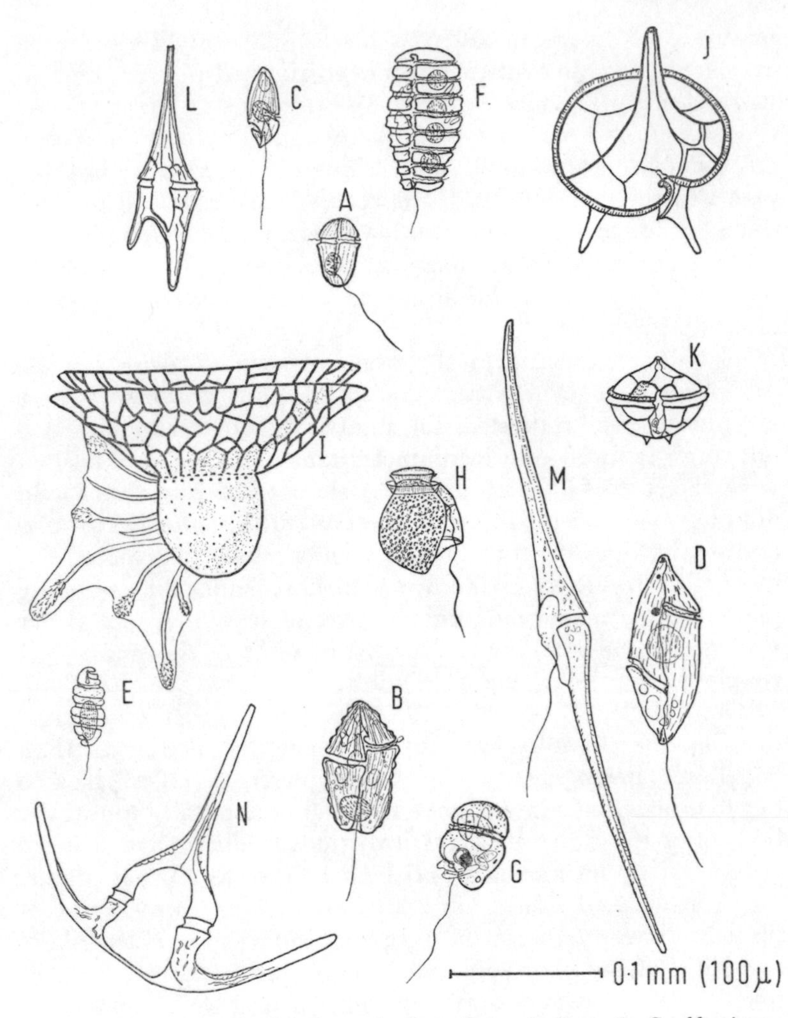

Fig. 10.—Planktonic dinoflagellates. A, *Gymnodinium hyalinum*; B, *G. abbreviatum*; C, *Gyrodinium glaucum*; D, *G. britannia*; E, *Cochlodinium brandti*; F, *Polykrikos schwarzi*; G, *Proterythropsis vigilans*; H, *Dinophysis acuta*; I, *Ornithocercus splendidus*; J, *Peridinium depressum*; K, *P. ovatum*; L, *Ceratium furca*; M, *C. fusus*; N, *C. tripos*. (*Most after Lebour*)

outgrowths which must make them much more resistant to sinking than the species found in colder water (Fig. 10: compare H, a temperate, with I, a tropical, form).

Methods of reproduction differ in these various groups of

plants. When a diatom divides the two halves of the shell are pushed apart as the cell enlarges, and eventually each new cell goes off with one half and forms another half which fits inside the old (Fig. 7). Thus one of each pair of daughter cells is smaller than the mother cell by just twice the thickness of the siliceous shell. Although this is a very minute fraction of the total, it does mean that in an actively reproducing population the cells get smaller and smaller and at last become too

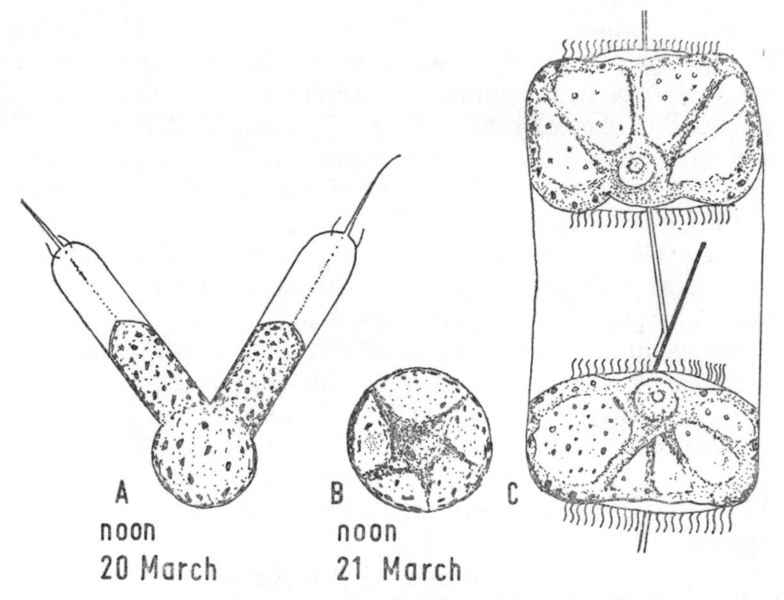

Fig. 11.—Auxospore formation in the diatom *Ditylum brightwellii.* A, formation of the auxospore in a narrow cell; B, completed auxospore; C, broad cells developed from an auxospore. (*After Gross*)

small to live. When this happens another kind of reproduction takes place, the formation of auxospores (Fig. 11). This is often a simple form of sexual reproduction and it takes place in different ways in different species. In sexual reproduction the nucleus divides so that the number of its chromosomes is halved, preparatory to fusing with a similar nucleus of the opposite sex, and thus restoring the full number of chromosomes. In diatoms there is some doubt whether there are two sexes. There may be in some diatoms (at least there are two sizes of reproductive cell), but in the planktonic species which have been investigated (and that is still only a few) the process of reproduction is very simple. The nucleus carries out a reduction division, then each daughter nucleus divides again in the ordinary way and of the four nuclei thus formed, two degenerate and the other two fuse with each other. By this time the cell body has slipped out of its siliceous shell and the fusion of the nuclei is the signal for it to increase greatly in size. It divides again, so does the nucleus, and a new set of frustules is formed, much wider than the old. This description applies to a pelagic diatom *Ditylum brightwellii*; in other species there are variations in the number of nuclear and cell divisions and the number of auxospores formed. The essentials are the reduction division and the subsequent fusion of two nuclei with the half-number of chromosomes, and the resulting large cell with the full chromosome number. The cycle can then start afresh.

Most observations have naturally been made in cultures where the diatoms are growing and dividing very actively, but auxospores have often been found in the sea. Indeed attempts have been made to identify particular patches of diatoms by measuring cell-diameters and by such repeated measurements to find the length of time between one auxospore generation and the next. For *Rhizosolenia styliformis* this has been estimated at 2½, and for other diatoms 1–5, years. This seems a very long time and more of such work is urgently needed.

The main cause governing auxospore formation is decreasing cell size, but auxospores may also result from unfavourable conditions. In *Ditylum* there is a size above which auxospore formation does not take place and a size below which the diatom cannot survive without it. Between these limits it seems to be controlled by external conditions. In poor light, or if nutrients are scarce, auxospores will be formed near the upper size limit; when light and nutrients are favourable, the diatom will go on dividing until the lower limit is reached. If no auxospores are then formed the cells disintegrate and die.

Vegetative reproduction and auxospore formation do not exhaust the possibilities for diatoms to multiply. There are also microspores. Diatoms of many species have been seen with the frustule filled with small bodies which look as if they had been derived from the diatom cell. There has been much argument and discussion about whether these are actually diatom spores or only the result of pathological divisions, or even parasites. However, it has been clearly shown that, in *Chaetoceros* at least, such spores are really formed by the division of the diatoms, are flagellated, and eventually escape to form new cells. There is a suggestion that this too may be a form of sexual reproduction but the nuclear history is not known.

Finally there is the resting spore, which is usually found when the environment becomes unfavourable; the cell sap is extruded and the plasma and nucleus contract into a small compact sphere in one corner of the frustule. Sometimes this is a mere temporary cessation of activity, as for instance in a culture of *Ditylum* in which most cells produced resting spores every evening which expanded to give ordinary cells every morning. In the sea, diatom resting spores may be found with thick and sometimes ornamented siliceous shells and these can sink to the bottom and survive over an unfavourable period which is often winter. Obviously only diatoms living in shallow water can make use of this kind of resting spore.

Flagellates usually reproduce by fission. The plane of division passes through the point where the flagella emerge ventrally, and is transverse or oblique. In the thecate dinoflagellates each daughter cell goes off with half the skeleton and a new half is formed. Since the plates of the skeleton meet edge to edge, there is no diminution of size with multiplication as there is in diatoms.

Sometimes the daughter cells do not use the old skeleton at all but slip out of it, each forming a completely new one. This happens in *Peridinium trochoideum.* To get out of its enclosing shell the plant secretes a little pad of hyaline substance, and the extra pressure this causes is enough to crack the shell along the horizontal furrow. The cell then escapes and divides, sometimes into two daughter cells of very unequal size. If the new shell is formed before there has been any more growth, this kind of division will account for the great range of sizes found in this and other species. In some species the daughter cells may remain joined to one another to form a short chain.

All these flagellates also form spores, sometimes resting spores which do not divide and sometimes spores which divide into four or more small cells. Whether there is ever a sexual process is not yet quite certain.

A little must be said here about methods of collecting phytoplankton and of estimating its abundance. An easy way is to tow slowly through the water a conical net (Fig. 18) which may be made of material with coarse or fine mesh; the finest bolting silk has 200 meshes to the inch (although a much finer mesh can now be had in nylon) and there are various grades of material from that up to meshes 1 or 2 mm in diameter. But the smallest organisms will slip through the meshes of even the finest net, and to estimate their numbers it is more usual to take a sample of water and either centrifuge it, or add preserving agent and wait till all the organisms have settled out. They can then be counted, either by examining the sedimentation tube from below with a reversed microscope or by removing the sediment to a ruled slide and counting either the whole or a definite fraction of the contents under a microscope. The difficulty is that no one method is suitable for the whole of the plankton. A combination of methods must therefore be used.

To count all the plants and animals from water samples or tow-nettings is a tedious and time-consuming job, so it is not surprising that there are comparatively few places where such counts have been made frequently and rigorously throughout a whole year. It is nevertheless useful to have observations even if they have been taken over only part of the year and even if they have been taken by methods which are not ideal. They will give us at any rate a rough estimate of the seasonal succession. Where counts have been made throughout the year, it has almost necessarily been on the coast; for open oceans we have until recently been dependent on the samples taken in the course of voyages by research ships. Now, however, the Weather Ships are making a contribution by collecting weekly plankton samples at their stations in the open oceans.

The seasonal cycle of phytoplankton abundance is very regular and in any one area occurs in much the same sequence year after year; but the times, the species, and the numbers naturally vary from one part of the world to another (see Fig. 12).

In the Arctic (Fig. 12(A)) only three seasons are detectable—

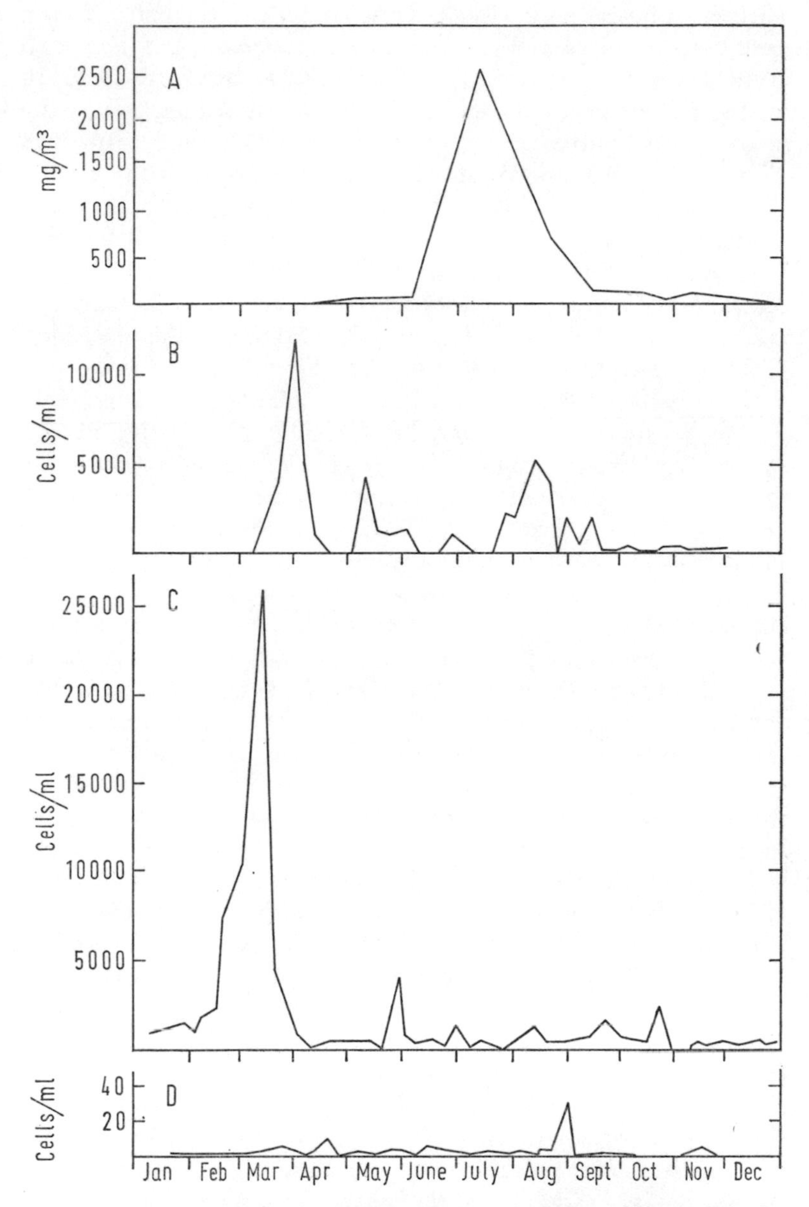

Fig. 12.—Seasonal variation in quantity of phytoplankton at different latitudes. A, Arctic (about 75 °N); B, Loch Striven (56 °N); C, Long Island Sound (41 °N); D, Great Barrier Reef Lagoon (16 °S); note that scales differ.

spring, summer, and winter. In winter the ice carries, frozen into its lower surface, some diatom resting-spores, but, although nutrients are abundant, there is little life below the ice. In spring the ice begins to melt, and round the edges and in the open cracks between the floes an astonishingly thick growth of diatoms appears and with them a few zooplankton larvae. In summer the zooplankton has grown up, is grazing down the diatoms and is reproducing. After this the numbers of plants and animals both decline gradually until the ice freezes up again. Nutrients are probably never lacking in arctic waters—it is the lack of light and of the vertical mixing of the sea which limits the multiplication of the diatoms.

This pattern, a single outburst of plant growth in early spring or summer, is found in Greenland and also in the Antarctic, but over most of the world there are two great increases, not one, and they come in spring and in autumn. In theory the spring increase is earlier and the autumn increase later as one moves towards the equator, until they coincide and there is a winter increase with perhaps one or more summer increases as well. In tropical waters the increases are less marked and the plankton is maintained at much the same level all the year round (Fig. 12(D)). In actual fact, however, there is so much variation from place to place, caused by topography or climatic conditions, that each region must be considered individually. Even so there are local variations from one year to another, both in time and size of increase and in the species occurring.

For the open ocean in northerly waters, a very good series of weekly samples was taken over a whole year (1948–49) from Weather Station M (66°N, 2°E) in the Norwegian Sea and another from Weather Station B in the Labrador Sea. In the first there was one big increase, consisting of dinoflagellates and coccolithophores as well as diatoms, lasting from May till August. In the autumn of 1948 (October to November), but not of 1949, there was a small increase, but this was made up of species carried north by the Atlantic current. At that time the water was mixed down to 75 m and it is rather surprising that the phytoplankton could multiply at all. The water in the open sea is however clearer than that near the coast, where there is much detritus in suspension, and photosynthesis can therefore go on into deeper water. The Labrador Sea (10° further south) showed a seasonal pattern of spring and autumn increase and summer minimum; it was unusual in that diatom and dinoflagellate numbers rose and fell together and that the phytoplankton was a mixture of species from many different areas. Station B was near the centre of a big eddy receiving water from the Greenland and Irminger currents, the Labrador current and even occasionally the North Atlantic Drift.

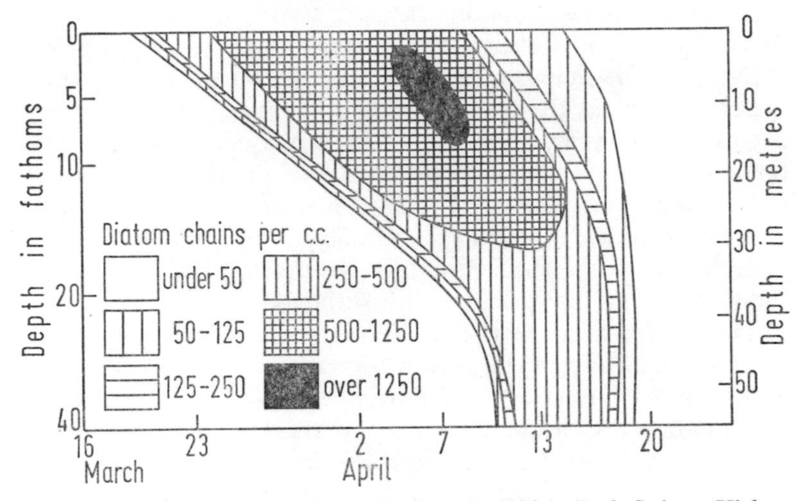

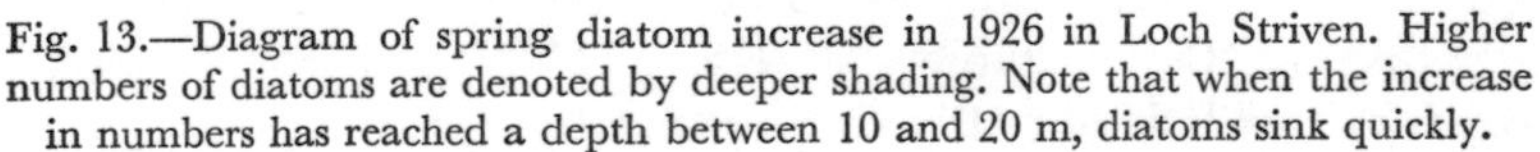

Fig. 13.—Diagram of spring diatom increase in 1926 in Loch Striven. Higher numbers of diatoms are denoted by deeper shading. Note that when the increase in numbers has reached a depth between 10 and 20 m, diatoms sink quickly.

The seasonal cycle of spring and autumn diatom increases is found in temperate waters all over the world (Fig. 12(B)). It is found on the coasts of Norway as far north as Tromsø, round the British Isles and on the French coast, on the east and west coasts of N. America, and in parts of the Pacific.

There is typically a winter minimum of plant life when nutrient salts are abundant but the sea water is much too stirred up to enable plants to remain in the photosynthetic zone. Some time in March or April, when the sea is getting stabilized (see p. 15), there is a sudden outburst of growth which lasts for a few weeks. As the diatoms multiply they use up first the nutrients in the surface water and then gradually those in deeper water (Fig. 13), until a depth is reached where there is not enough light to sustain photosynthesis. When this level is reached the increase is over and the diatoms sink down towards the bottom.

Summer is a period of phytoplankton scarcity except in water which is turbulent (the Straits of Juan de Fuca, south of Vancouver Island, for instance) or where storms may stir up the deep water which is still rich in nutrient salts (e.g. Loch Striven, Firth of Clyde). In such areas there may be a succession of small phytoplankton increases throughout the summer. Since the zooplankton is by this time numerous and grazing actively on the diatoms, these summer increases do not reach such a density as the spring one. The onset of the second big increase of the year, the autumn one, is very variable and may be as early as August or as late as October. It usually comes when colder weather is making the water column unstable again, so that the nutrient salts are beginning once more to be distributed evenly throughout the water column, and when the light below the surface is still sufficient for photosynthesis. Lack of light and instability of the water are probably the factors which bring the autumn increase to an end. After this plant life declines until the following spring.

Although this account applies mostly to diatoms, dinoflagellates and other groups have their seasons of abundance too. Dinoflagellates very often follow diatoms, increasing in numbers as the diatoms die away. They are usually most abundant in late summer and autumn.

As already mentioned, the disposition of land and water in any particular place may make a big difference to its plankton. In the channels between the islands of the San Juan Archipelago near Vancouver (49° 45′N, 132°W) the water is always mixed because of strong currents coming in from the Pacific. The spring increase starts late, and from May till September there is a series of diatom outbursts with one species succeeding another. Quite close by, but in long sheltered inlets where the water is stabilized early in the year, the spring increase starts in March and there is a succession of increases till July.

Although along the northern part of the American east coast there is this spring and autumn diatom cycle, the situation changes south of Cape Cod. In the shoreward areas even of Cape Cod Bay and the Gulf of Maine diatoms make an early appearance, and in the shallow coastal waters of Woods Hole, Narragansett Bay and Long Island Sound (41–42 °N, 70–74 °W) there is a winter diatom maximum and a series of increases in summer either larger or smaller than the winter one (Fig. 12(C)). The same pattern is found in the Adriatic (43° 30′N, 16° 30′E) near Split.

Off the coast of California (33 °N, 117 °W), where there is frequent up-welling of deep water carrying nutrients, the phytoplankton is most often at its maximum during the second quarter of the year, but increases may come at any season. The actual numbers of diatoms and dinoflagellates recorded were very low, but this is probably partly because, although large samples of water were drawn, they were subsequently concentrated by pouring through a fine silk net which failed to retain the smallest plants.

Not much continuous work has been done in tropical waters. In the lagoon of the Great Barrier Reef (16 °S, 146 °E) the quantity of the phytoplankton at any time was very low and the few small increases which occurred seemed to come after storms which had perhaps caused a stirring up of nutrients from the bottom (Fig. 12(D)).

Much work has been done in the Antarctic by the R.R.S. 'Discovery' and other ships investigating whaling. North of the Antarctic circle there is no land to interrupt the vastness of the southern oceans, and their waters stretch round the world. There is little difference therefore in the phytoplankton species present or the seasonal cycle in any area except in a north and south direction. Although South Georgia is the same distance from the equator as southern Scotland, the influence of the ice reaches further towards the equator in the southern than it does in the northern hemisphere; the seasonal pattern off South Georgia approaches that of the Arctic, i.e. a single midsummer diatom increase, with only a slight break in the gradual decline which follows, to correspond to an autumn maximum. Further south still, even that disappears and there is only a single peak of phytoplankton growth. Diatoms predominate in the phytoplankton community. The edge of the ice pack acts as does the land, and the plants are richest there and round the islands. Since nutrient salts are never lacking in antarctic waters one may wonder why the diatom increase lasts no longer than it does. It may be partly because silica, used for making the diatom frustules, is limiting and partly because of the rich zooplankton which develops in summer and grazes down the phytoplankton.

Although there are many hundreds of species of diatoms and dinoflagellates, any one of which, given the right conditions, could probably multiply enormously and dominate the plankton, yet certain species are perennially important and in their season become the dominant species year after year. Among these is *Skeletonema costatum*, a small pillbox-like form of 4–16μ in diameter, whose cells are united in long chains by delicate siliceous rods. It is a neritic diatom and flourishes off the coasts of the north Atlantic and Pacific. In the open ocean, both in the Arctic and Antarctic, *Fragilaria* and related species of pennate diatoms are more common. In warmer water the numerous species of *Chaetoceros* and *Rhizosolenia* are more important. In the tropics the proportion of dinoflagellates and coccolithophores increases and this is also true of the open oceans.

The importance of the minute flagellates in the economy of the sea is not yet known. They are too small and too delicate to be counted in water samples in the usual way, and although enormous numbers have been reported in some areas little is known about their seasonal variation.

During a diatom increase one species is usually predominant, but two or three species may be abundant at the same time or succeeding one another. The same species or the same succession may occur year after year, but in some years a diatom or dinoflagellate which is usually scarce will appear in enormous numbers. The causes of these variations are not yet known, but different species probably have different requirements in conditions of light, temperature, nutrients, trace elements or vitamins. For instance it has been found in Long Island Sound, where *Skeletonema costatum* or *Thalassiosira nordenskioldii* dominates the spring increase in different years, that at low temperatures and low light intensity the latter grows better than the former. Apparently therefore it is the temperature of the water in winter or early spring that determines which species is going to predominate. Dinoflagellates are said to be able to thrive with a lower nutrient concentration than diatoms, which is probably why they usually follow diatoms.

Sometimes a phytoplankton organism will multiply to such an extent that the water is coloured by it. This happens most often with dinoflagellates but is known with other groups too. These concentrations may be harmless in themselves, although they may lead to unpleasant consequences.

The "Red Tide" or "Red Water" which occurs off the coasts of California, Florida, S.W. Africa, India, and elsewhere is usually caused by enormous concentrations of dinoflagellates and these often have a catastrophic effect on fish and indeed on marine life in general. Dead and dying fish and crabs may be thrown up on the shore in thousands. Death may be caused simply by suffocation but some dinoflagellates are actually toxic and these can be dangerous to man also. During the season when red water is likely to occur, the taking of shellfish is forbidden on parts of the Californian coast because oysters and clams that have fed on the dinoflagellates, although they do not suffer themselves, cause paralytic poisoning and even death to those who eat them. On the coasts of Florida such red tides have been frequent in recent years and, besides killing fish and crabs, they affect human beings by what is described as an "irritant gas" causing choking and coughing. This is more probably due to small air-borne particles of the toxic substance, for the effects are felt mainly when there is an onshore breeze.

In India, on the Malabar coast, the same thing happens but, although fish are killed in large numbers and the rotting bodies give rise to an unbearable stench, no human illness has been reported. Such outbreaks have been known for centuries, and could provide an explanation for the miracle which saved the Portuguese garrison of Cannanore in 1507. The besieged were reduced to eating rats and lizards when on August 15th they prayed to the Virgin for help and suddenly the shore was strewn with dead fish and crabs. In that area red water usually comes near the end of August, and since the 1507 date was reckoned on the old style calendar ten days must be added, which brings it to just about the right time.

Besides dinoflagellates, blue-green algae and various flagellates can cause "water blooms" of different colours, red, yellow, brown, blue or green, and some of these too may be poisonous. It is a curious fact that although a bloom of the small flagellate *Prymnesium parvum* caused mortality of fish in a fish pond, it proved non-toxic when grown in pure (i.e. bacteria-free) culture.

A great deal remains to be found out about these death-dealing blooms and their causative organisms. Some of the conditions favouring their growth seem to be abundant nutrients (caused by up-welling or by the death of a previous abundant plankton), a calm warm sea, and sometimes low salinity. They are much commoner and usually more noxious on semi-tropical or tropical coasts than in temperate waters.

There has been much discussion about the best way of estimating the amount of phytoplankton present. The method which gives most information is undoubtedly to take large water samples and count the individual cells contained in them, but the numbers themselves do not mean very much unless we know the component species. It has been calculated that 25,000 cells of *Fragilaria nana* would be needed to make up the volume of one *Coscinodiscus centralis* cell, so it is obvious that the size of the different species must be taken into account. At the height of the main increase the numbers in different places may vary from a few score to a few thousand cells per millilitre. Apart from chemical and physical factors some of this variation will be caused by differences in methods of sampling, but some by differences in the needs of the organisms themselves.

There are difficulties too in that the size of the samples is very small compared with the extent of the water mass. An assumption must therefore be made that the phytoplankton is evenly distributed within the water mass to be sampled; this is certainly not always true. A method of continuous sampling as the ship moves along (like that of the plankton recorder for zooplankton (see p. 56)) would be very useful. Such an instrument has been designed and used in California but the results have not yet been published.

A way of overcoming the counting (although not the sampling) difficulties is to filter a water sample through a membrane filter (the finest of these retain particles down to a size of less than 1μ), extract the chlorophyll and measure this. Since the amount of chlorophyll depends on the amount of photosynthesizing tissue this method gets round the problem of different cell sizes, but it is not entirely satisfactory. The more delicate cells break up on the filter, letting some of their chlorophyll through; in addition, there are several types of chlorophyll, not all of which are equally important in photosynthesis. A measurement of the silica present in a water sample has been used to estimate diatoms only. It is obvious that there is as yet no completely satisfactory way of measuring the amount of active plant tissue present, and that at present a combination of methods is necessary.

Even if we could be sure of getting an accurate estimate of the number and kind of plant cells or of the amount of active photosynthesizing tissue present at any one time we should still not know enough to measure the productivity. What we had measured would have been only the standing crop, the amount present at one particular moment, and to get an idea of the actual plant production one must take into account the speed of reproduction and the amount which has been eaten by animals. If a diatom increase lasts longer in spring than in summer it is partly because in summer there are more animals present to eat it up. Owing to the higher temperature a life cycle is passed through more rapidly in the tropics than in temperate regions and so there is a more rapid turnover, and production, as indicated by the numbers present at any time, appears less than it actually is. The converse is true in arctic and antarctic waters. We can get at this problem in another way, by measuring the amount of some nutrient, such as phosphorus, taken out of a particular area during the course of a year and calculating, from our knowledge of the average phosphorus content of the phytoplankton, the production which has taken place. These methods are dealt with in Chapter 3.

Chapter 3

Primary Production

> O what an endless work have I in hand
> To count the sea's abundant progeny
> Whose fruitfulle seede farre passeth those in land
> And also those that wonne in th'azure sky
> For much more eath to tell the starres on hy
> Albe they endlesse seeme in estimation
> Then to recount the sea's posterity,
> So fertile be the flouds in generation,
> So huge their numbers and so numberlesse their nation.
>
> EDMUND SPENSER, "The Faerie Queene"

Since the phytoplankton is the basis for all life in the sea it is important to know how it varies in different areas. To count the phytoplankton cells tells us only what is there at the moment, but to estimate the productivity we have also to know the rate of reproduction. A count of cells present in the sea does not tell us much if the plants are being grazed down by plankton animals as fast as they are produced, or if breakdown of dead plants by bacteria is going on actively at the same time as photosynthesis.

If we compare a temperate and a tropical region the first will usually be found much the richer in spring and summer, but in the tropics the high temperature speeds up reproduction and the total number of cells produced over the whole year may be as great as in temperate seas.

There are several factors which affect the productivity and of these the most important are light, temperature and the supply of nutrient salts.

The depth to which light penetrates in the sea has often been measured. In 1891 Regnard germinated seeds of cress and radish at different depths in the Mediterranean off

Monaco and found that at 30 m hardly any chlorophyll was produced. Nowadays there are instruments which can be sunk to any depth and will measure not only the quantity but the quality of light there. Visible light penetrates more deeply into the sea than that at either end of the spectrum. The infra-red and ultra-violet rays are absorbed within the first metre. Most

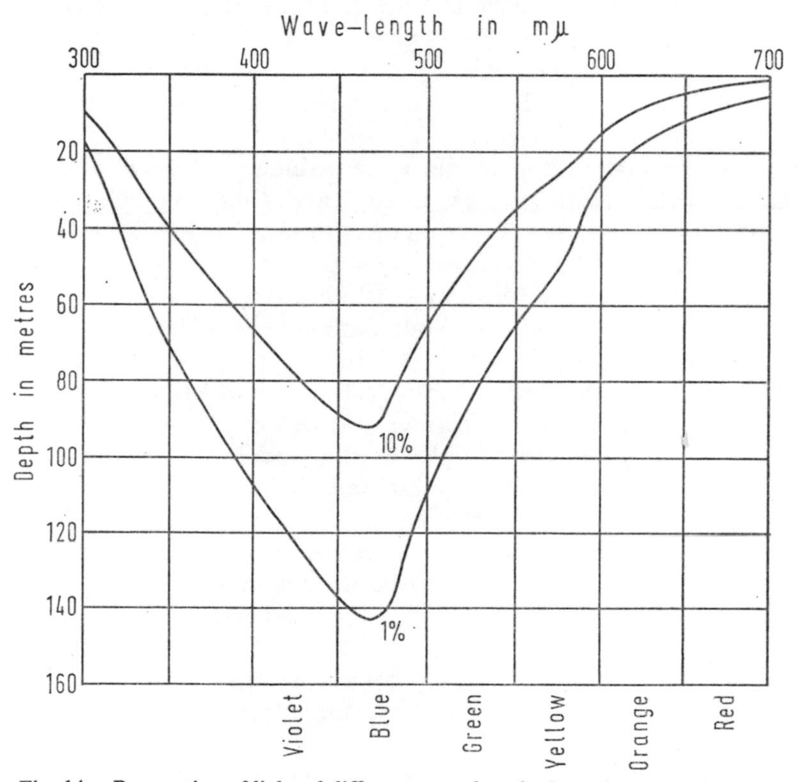

Fig. 14.—Penetration of light of different wave-lengths into clear oceanic water. (*After Jerlov*)

photosynthesis takes place in the blue to yellow range and these are the rays which reach deepest into the sea (Fig. 14), the blue rays penetrating farthest in clear oceanic water and the yellow-green rays near the coast. The amount of light does however decrease fairly rapidly with depth: a day lasting eleven hours at the surface will last for seven hours at 20 m, five hours at 30 m, and only a few minutes at 40 m.

In most places, and over the oceans as a whole, the light near the surface throughout the year is probably enough to enable the phytoplankton to deal with all the nutrient salts available, but there are certainly areas and seasons where lack of light prevents growth. The Arctic and Antarctic oceans are examples; during the continuous winter night there can of course be no photosynthesis, and in spring and autumn there the shortness of the day limits production. Even in the summer the nutrient salts are not all used up and this may be because the oblique rays of the sun do not penetrate deeply into the water.

The vertical mixing of the water which goes on in winter in temperate latitudes (see p. 15) carries diatoms below the depth at which they can photosynthesize. There is a light intensity, called the compensation point, where the organic matter used up in respiration just balances the organic matter produced by photosynthesis and, if mixing goes below the depth at which this light intensity is found, spring growth cannot start. The critical depth is that depth above which total photosynthesis equals total respiration and this can be calculated if we know the amount of light at the surface, the amount of light at the compensation point and the transparency of the water. It is greater than the depth of the compensation point. Sverdrup found that in the Norwegian Sea in 1949 (Fig. 15) the critical depth was close to the surface in early March and gradually went deeper as the spring days lengthened. In early April, for the first time, mixing did not go below the critical depth and a diatom increase began. Stormy weather interrupted it but from May onwards a stable layer was formed and diatoms flourished.

In general the temperature of the sea ranges from −1·9 °C to about 25 °C. Chemical changes go on more rapidly at higher temperatures, and a rough guide is that a rise of 10 °C will double the rate of a reaction. Thus in tropical seas we should expect plant growth (and also the breakdown of plants by bacteria) to be more than four times as fast as in polar seas. The main importance of temperature lies however in its effect on the stability of the sea.

It has been said that the replenishment of the nutrient salts in the productive surface layers is the essential factor determining the magnitude of organic production. The amount of the nutrients is constantly varying according to their depletion by the photosynthesis of the phytoplankton, the excretion by the zooplankton, the bacterial breakdown of both phyto- and zooplankton and, not least, by water movements.

We have seen in Chapter 1 that nutrients are on the whole abundant in the surface water at the poles and poor in the tropics because of the stability of the water in warm seas. In

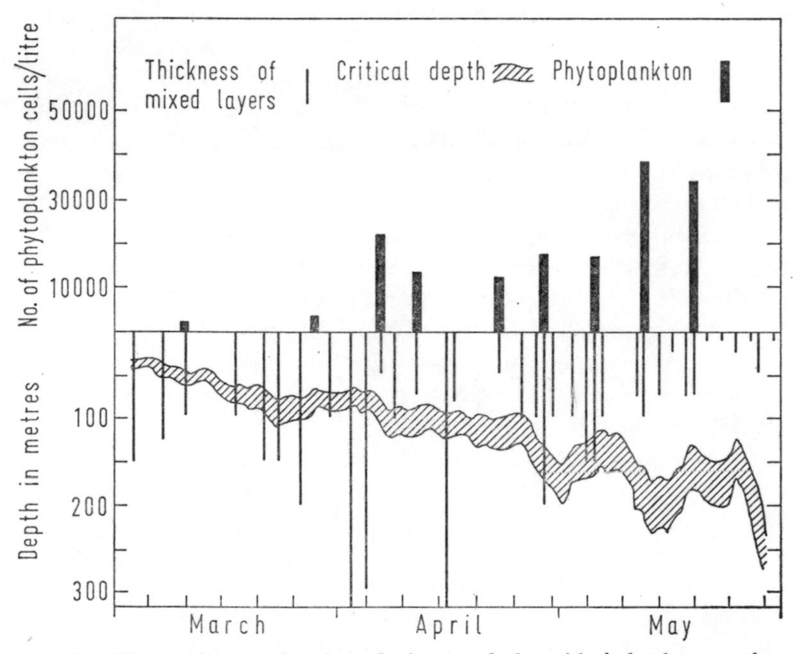

Fig. 15.—Observations on the phytoplankton and the critical depth at weather ship M in the Norwegian Sea in the spring of 1949. (*Adapted from Sverdrup*)

temperate waters the seasonal changes in temperature and the vertical mixing which follows them ensure an annual supply of nutrients to the surface. There are also places where a deep nutrient-rich current strikes against an underwater slope and is brought to the surface. This up-welling happens off the west coasts of Africa and the Americas and leads to an extraordinary abundance of plankton and fish. It happens too where there are strong tides over a rough bottom, as in Puget Sound, Washington (see p. 36) or the Bay of Fundy, Nova Scotia.

We have seen that nitrate and phosphate are the most important nutrient salts and each has been found at some time to be a "limiting factor" (i.e. its absence or scarcity sets a limit to plant growth). But these are not the only salts necessary. Silica may be limiting in the Antarctic, where phosphate and nitrate are always abundant. In temperate waters too the spring diatom increase often dies down before these salts are quite exhausted (sometimes because of zooplankton grazing) and it is possible that something else, perhaps some trace element, is then limiting growth. The succession of one diatom species by another or by another type of phytoplankton may also be caused by the differing requirements of different species, or even by harmful products of the metabolism of the plants themselves.

The productivity of an area may be measured in a number of ways. If it were done by assessing the reproductive rate of the phytoplankton, counts would have to be made frequently of all the organisms, plant and animal, and one would also have to take account of their size. This is now possible by using a mechanical counter in which the numbers of particles of different sizes can be estimated, but no such phytoplankton studies have yet been published.

A more realistic assessment could be made if we knew the total weight or, better still, the total organic matter of the phytoplankton. In recent years (see p. 51) an attempt has been made to estimate this by measuring the chlorophyll used in photosynthesis. It is a fairly simple estimation but it has its drawbacks in that the relationship between the chlorophyll and the weight of the phytoplankton cell is not constant, and chlorophyll from dead phytoplankton may be included in the estimation. However, it is a useful guide. A measurement of carbohydrate or organic matter is also useful but suffers from some of the same drawbacks.

Another method of measuring productivity is to estimate the total amount of some nutrient which has been used up during the season. This has been done in the English Channel, for phosphate, nitrate and silica. Winter and early spring is the time when the nutrients are at their richest and are evenly distributed from surface to bottom (Fig. 5), so that it is easy to calculate the total amount under 1 sq. m of surface. After a stable layer has formed and the surface has been depleted of nutrients the measurement is repeated for both layers and, since we know the amount of nitrate, phosphate, and silicate which a diatom contains, we can calculate what weight of diatoms has been produced. The nitrate and phosphate calculations gave results consistent with one another; the silica did not. This may be because some of the photosynthetic organisms were not diatoms and did not use silica, or else—what is more likely—the silica of the dead diatoms had dissolved quickly and had been used again immediately. This regeneration and recycling can happen with other nutrients too and is something which throws the calculations out, making the results too small.

The excretions of the zooplankton also contain nitrogen, phosphorus and other substances which can be used by phytoplankton and it has only recently been realized how important a part they may play even in surface waters. In Long Island Sound it was thought that during active phytoplankton growth in summer half the nitrogen used came from zooplankton excretion. An advantage to the plants is that this is a quick process; bacterial decay is much slower.

A more accurate way of measuring productivity is to measure photosynthesis. We can find out what the plants in the sea are actually producing by putting a sample of the sea water in a bottle and exposing the bottle, in the place where the sample was collected, for a fixed time, usually a day or half a day. Diatom cultures have been used for these experiments, as well as natural sea water. The plants produce oxygen and from the increase in this we can calculate how much organic matter they have built up. We must correct for their respiration and this is done by measuring the decrease in oxygen in a similar bottle kept in the dark and adding this loss to the increase in the bottle kept in the light. It is a most useful method of testing production in different depths and places, and when it is done over the course of a year we can get a picture of the production of an area.

It has been found that photosynthesis increases with increasing light intensity only up to a certain point; above that, light injures the plant cells and their activity falls off. In summer in temperate latitudes the maximum production is therefore found, not at the surface, but at some metres below, and not

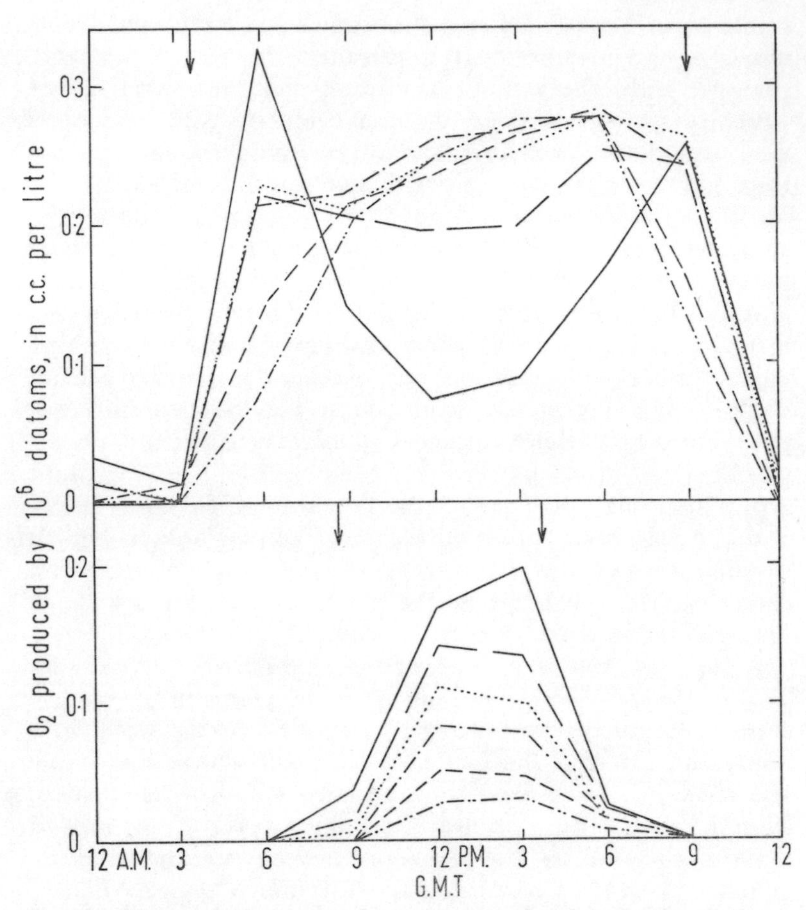

Fig. 16.—Photosynthesis (measured as oxygen produced in bottles of diatom culture) at different depths during a sunny summer day (upper figure) and a dull winter day (lower figure). Arrows mark sunrise and sunset.

——— surface, —— —— ½ metre, ... 1 m, —— · —— · —— 2 m,
— — — — 4 m, —— ·· —— ·· —— ·· 6 m.

at midday, but in the early morning and evening (Fig. 16). The compensation depth is found quite near the surface on dull winter days but it moves deeper as the year goes on and reaches from 20–40 m in summer. This is for temperate latitudes and coastal water. In clear oceanic waters, such as those of the Sargasso Sea, it may be as deep as 120 or 150 m. In temperate latitudes the great difference between winter and summer is due not so much to the increased intensity of the light as to the greater depth to which light penetrates in summer, and the much longer day.

On the whole production by photosynthesis in the sea is not very efficient. It has been variously calculated as using from less than 0·1% to 20% of the energy of visible light under ideal conditions. This inefficiency is caused mostly by the reflection of light from the surface and its rapid absorption by sea water.

Measurement of photosynthesis has recently been made much quicker and very much more delicate by the use of radioactive carbon, ^{14}C, as a label for the carbon dioxide used by the phytoplankton. A little radioactive carbonate is added to each bottle of sea water, and the amount taken up by the plants in a day or half a day is measured after exposure in the sea or, more simply, after exposure to a standard light source. This method has been used along with the "light and dark bottle" method (measuring oxygen) or sometimes alone. Since there is little time to spare on oceanic expeditions the bottles have usually been exposed to standard light on board ship so as not to delay the voyage by spending 12 or 24 hours at one spot.

In areas where production is high the two methods agree well but where it is poor, as in tropical oceans, they may differ considerably, the radiocarbon method usually giving lower results. There has been much argument about the cause. It seems that some of the carbon taken in by the plant for the photosynthetic process is immediately used for respiration and excreted again, so that what the radiocarbon method measures is something between total photosynthesis and net photosynthesis (i.e. photosynthesis minus respiration). It has been suggested, on the other hand, that respiration in the dark bottle of the oxygen estimation method is enhanced by the respiration of bacteria which may not survive in the bottle exposed to light. Since the respiration figure comes into the calculation for total photosynthesis it would make this too high. The question of respiration is very important for, unless it bears a constant relation to photosynthesis, calculations based only on production in the light will not be very reliable. While a phytoplankton cell is healthy and reproducing actively the respiration does seem to stay rather constant, at about 10% of oxygen production. When the cell is unhealthy, however, or when it is short of nutrient salts, the respiration figure may rise to equal that for photosynthesis.

As so often happens no one method is ideal; each has its drawbacks, but imperfect as they are each has greatly enlarged our knowledge of the processes going on in the sea.

It is interesting to make an estimate of the total production over a year and this has been done for quite a number of regions. It is more of a guess than a real estimate, because of the uncertainties of the methods and the great variations from year to year. The figures vary from about 40–60 g of carbon for every sq. m of surface (i.e. for a column of water 1 m square from surface to bottom), which can be translated into about a quarter of a pound of starchy material per sq. m per annum. This may not seem much until we remember the enormous area of ocean and the millions of square metres it covers. However, the figure is a little below that of a square metre of arable land surface.

With the growing populations of the world and the increasing difficulties of feeding them, it has occurred to many scientists that it would be a good idea to grow photosynthetic organisms in culture and supply them as a dried food. Large-scale experiments were made in Japan, Israel and the United States and a pilot plant was set up in Massachusetts. The organism chosen was *Chlorella*, a one-celled microscopic green alga. The vessels in which the algae grew varied greatly—flasks, horizontal and vertical tubes, and a concrete trough were among them. In the pilot plant ingenious use was made of a four-foot wide plastic tube. This material was cheap and when filled with culture medium took up a flattened oval shape with an air space which could be filled with carbon dioxide. Since the object was to get the greatest possible yield, nutrients and carbon dioxide were supplied in excess. The culture was so thick that light was absorbed within a few inches and it was therefore kept stirred. This ensured that every cell was exposed to the sunlight but not for long enough to injure it. It had been found that intermittent light, with the cell exposed for only one-tenth of the time, gave optimum growth.

The yield varied from about 2–12 g (about 1/3 oz) per sq. m

per day but it was thought that in a better climate and under ideal conditions it might rise to nearly an ounce.

In some ways the results were encouraging; the product had a high protein content and when dried proved suitable as part of the diet of rats and chickens. It had a vegetable flavour and could be incorporated in human diet too, although its main use would probably be as an animal feeding stuff. However, it was expensive to produce and its use would at present be uneconomic. This is probably why so little has developed from these interesting experiments.

Since every year in temperate latitudes the phytoplankton cycle comes to an end in winter and starts all over again in spring it is obvious that the processes of building up and of destruction must balance.

The photosynthetic process is the chief means of building up living phytoplankton material and it is on one side of the equation; on the other side are all the factors that tend to limit growth or destroy plant cells. Among these are respiration, variations in light intensity, lack of nutrient salts, vertical movements of the water, sinking of dead cells, and the grazing of the zooplankton on the phytoplankton. If we could fit values to all these factors we might, by measuring a phytoplankton population and some of the variables, be able to predict what was going to happen to it.

This is just what a few mathematically-minded scientists have been doing. They have given to these factors values which are founded on research work done in the field or in the laboratory, and have set up mathematical models to represent what goes on in the sea. To do this everything must be very much simplified, and various assumptions made. Thus it would be impossible to consider separately all the species and sizes which make up a phytoplankton population; it must be represented by something proportional to the total mass of living material and for this chlorophyll is usually taken. As we have seen it is not a very accurate measure but at present it is the best available. Then we must assume that under ideal conditions photosynthesis goes on at a constant rate and that respiration bears a constant relation to it.

The amount of light falling on the sea surface and the depth to which it penetrates throughout the year have been worked out for different latitudes and these calculations are used to modify the photosynthetic rate. The phytoplankton is assumed to use up nutrients at a constant rate, and one or more nutrient salts can be measured in the sea. Usually for simplicity only phosphate or sometimes both phosphate and nitrate are measured, but as we already know these are not the only nutrients affecting plant growth.

From measurements of salinity and temperature the amount of vertical mixing going on can be calculated; and from

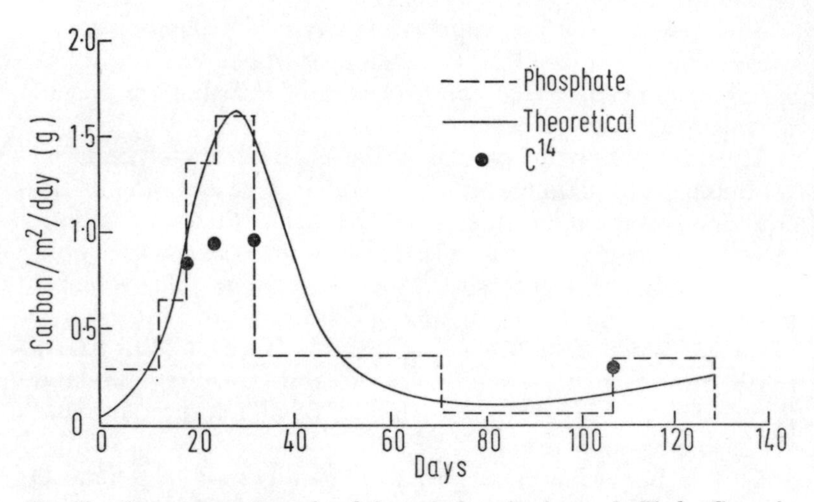

Fig. 17.—Comparison of actual and theoretical production on the Fladen Ground, North Sea, in 1956. (*From Steele*)

laboratory experiments the rate of sinking of some dead diatom cells is known, so that a figure can be put for the loss of phytoplankton below the photosynthetic level. Zooplankton can be counted and is assumed to graze at a constant rate. In this way a study can be made of an actual situation in the sea and the spring diatom increase has been thus studied on Georges Bank in the Bay of Fundy, and on the Fladen Ground in the North Sea. Surprisingly enough, in spite of all the assumptions and simplifications the course of the increase, as derived from the mathematical equations, agreed not too badly with what actually happened (Fig. 17).

This probably means that the factors we have been considering really are the most important in influencing what goes on.

As our knowledge grows and we can get more accurate figures for, say, the living matter in a phytoplankton population or the rate of grazing of different zooplankton animals, the mathematical model will come closer to the real situation and we shall understand better what is going on and how variations in the different factors can alter the course of events. But to do this much more research work is needed both at sea and in the laboratory.

Chapter 4

Zooplankton

> And I replied unto all those things which encompass the door of my flesh, "Ye have told me of my God, that ye are not He; tell me something of Him." And they cried, all with a great voice, "He made us." My questioning them was my mind's desire and their Beauty was their answer.
>
> ST. AUGUSTINE, *Confessions*

The zooplankton is the next step in the food chain which leads from the phytoplankton to man. Most of the zooplankton animals are therefore larger and more active than the plants, although their motive power is still feeble and their distribution depends mainly on tides and currents.

For catching zooplankton the use of a simple tow-net is still one of the best ways if we do not need to know exact quantities (Fig. 18). The method is open to several objections, namely that the smallest organisms will pass through the meshes, the largest may move so quickly that they can escape the net, and we do not know how much water is filtered. There are also difficulties when we want hauls from deep water. These drawbacks can be circumvented in one way or another. Almost all zooplankton organisms will be caught by using the finest mesh (down to an aperture of 61μ) but the sample taken of large and active organisms will be too small to be reliable. For this reason a thorough sampling of any body of water can be made only by using several nets of different mesh-size. There are numerous devices for opening and closing tow-nets while towing either vertically or horizontally and by using these we can find out which animals live at what depths. Finally, a flowmeter can be fixed in the mouth of the net to measure the quantity of water which has passed through.

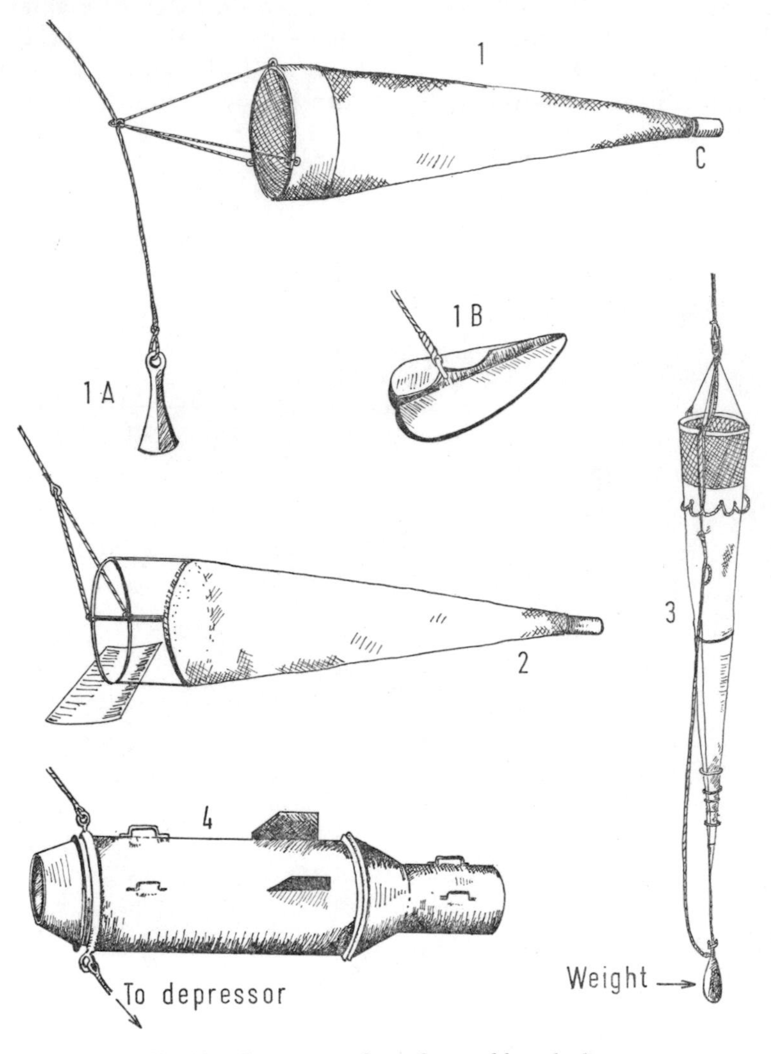

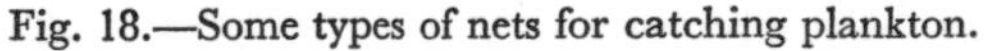

Fig. 18.—Some types of nets for catching plankton.

1. The simple tow-net: it can be kept below the surface by a weight (a) or a depressor (b); the bucket (c) retains the catch in good condition.
2. The paravane net: the diving plane acts as a depressor and the opening is not obstructed by bridles.
3. The Nansen net, rigged for vertical hauls, as used by the research ship "Discovery".
4. The Gulf III high-speed plankton sampler. (*From Fraser*)

Recent developments have been in the direction of making high-speed tow-nets, sometimes made of metal (Fig. 18(4), Gulf III net), with a reduced opening to deal with the greatly increased rate of flow. Such nets can be towed from a travelling vessel.

The Hardy plankton recorder (Fig. 19) is an instrument,

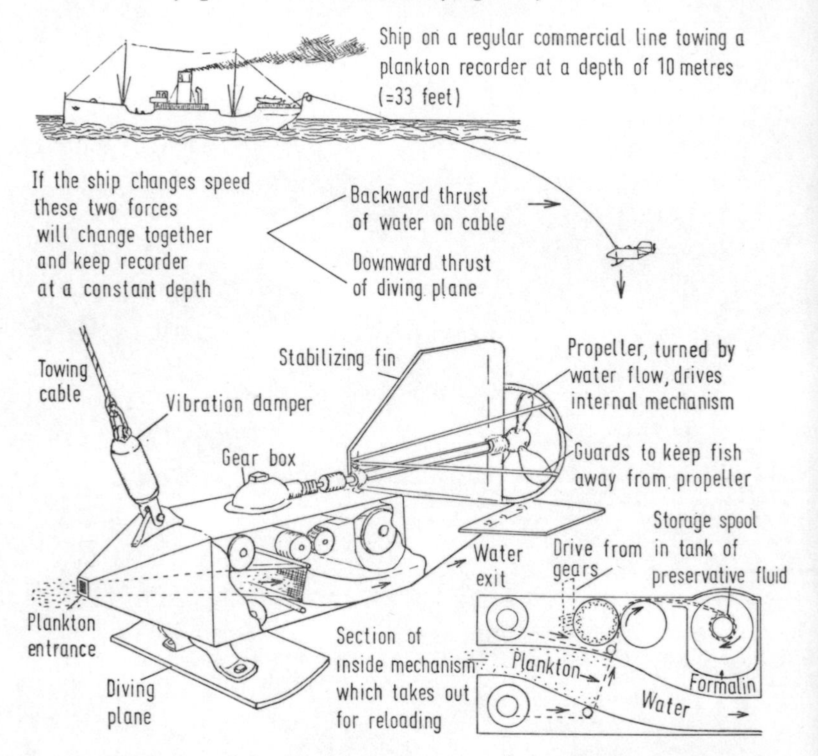

Fig. 19.—Diagrams of the Hardy plankton recorder, explaining how it works. (*From Hardy*)

now in use over much of the North Sea and North Atlantic, which has greatly enlarged our knowledge of the distribution and seasonal variation of the zooplankton. It is designed for towing at a constant depth from a commercial ship in passage and is therefore simple in operation and robust in construction. Water passes through a small opening at the front end, an opening which expands into a tunnel so that the water is slowed down to pass through a moving band of filtering silk.

This band is moved by gears attached to a propeller fixed to the outside of the instrument which is turned by the water flowing past. The propeller's speed of movement can be regulated by adjusting the blades and it is set so that one inch of silk represents a definite distance travelled. As the silk winds on after collecting its plankton it is covered by a similar band of silk and the whole is wound up into a tank of preservative. Thus a voyage of many miles can be represented on a band of silk tow-netting, and the animals can be recognized and counted with an exact knowledge of the spot where they were caught.

Zooplankton can be divided into two categories: those which are planktonic throughout their lives and are quite independent of the bottom, and those which are stages in the life history of the shore- or bottom-living animals and spend only part of their lives in the plankton. The second are naturally found in waters fairly near the coast and rarely in the open ocean.

Competition on the shore is fierce; on some rocks, for instance, you would not think there was room for one more barnacle or mussel and yet every spring millions of larvae are produced and drift off, the majority to perish but a few no doubt to find a settling place and grow up.

Planktonic larvae are often fantastically different from their parents (Fig. 20 and Plate III) and it is understandable that when they were first discovered they were described as new forms and given names of their own. This is why we speak of the Zoea larva of the crab, the Pluteus larva of a sea urchin (Fig. 20(A,B)), the Cyphonautes larva of a polyzoan (Fig. 20(F)), or the Actinotrocha larva of a Phoronid worm (Fig. 20(E)).

Many larvae when first hatched from the egg have bands of cilia round their bodies which by their constant beating keep the larva from sinking and enable it to swim actively. As it grows these bands enlarge too and are often carried out on projections of the body. The increasing number of cilia have of course to support an increasing weight. In sea urchin larvae the ciliated bands are carried out on long arms which are strengthened by skeletal rods (Fig. 20(A)); in starfish and Phoronid larvae the arms remain flexible (Fig. 20(C–E)).

Larvae such as these are so different from the adult that a

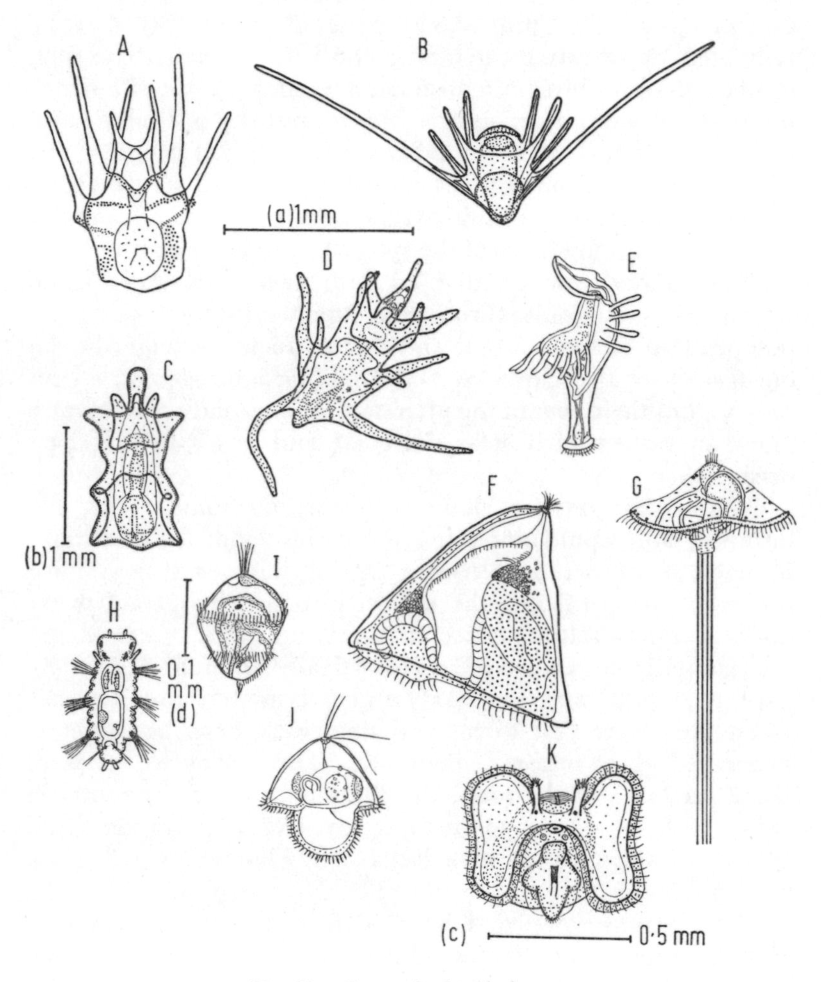

Fig. 20.—Some planktonic larvae.

A, Echinopluteus larva of sea urchin; B, Ophiopluteus larva of brittle star; C, Bipinnaria larva of starfish; D, Brachiolaria larva of starfish (later stage); E, Actinotrocha larva of *Phoronis*; F, Cyphonautes larva of polyzoan; G, Mitraria larva of the polychaete worm *Owenia*; H, Nectochaete larva of the polychaete *Nereis*; I, Trochophore larva of a serpulid polychaete; J, Pilidium larva of nemertine worm; K, veliger larva of gastropod *Nassarius*.

A and B to scale (a); C and D to scale (b); E, F, G, H, J, K, to scale (c); I to scale (d).

metamorphosis is necessary for one to change into the other. As the time for this change draws near the larva, getting bigger and heavier, sinks into water near the bottom and finally attaches itself. Many larvae have some power of choosing; if they settle first on an unsuitable ground they can cast off and continue for a few more days of pelagic life, drifting about and trying out different grounds. Eventually they attach themselves and then metamorphosis happens quickly. The larval arms are cast off and the larva takes on the adult shape.

Polychaete worms, at least those with planktonic larvae, are hatched with two rings of cilia but they quickly develop appendages and bristles like those of the adult, the cilia are lost, and the larva grows into the adult without any marked change. The same is true of molluscs. Both snails and bivalves maintain themselves in the plankton by a ring of cilia, the velum, which gets bigger as the larva gradually grows into the adult form, but is eventually shed.

Crustacea have no cilia and their larvae, although they may look very different from the adult, are still recognizably crustacean. Several groups hatch as an oval body called a nauplius (Fig. 21(J,M)), having three pairs of limbs and a single median eye. This gradually acquires more pairs of limbs, more body segments and usually a segmented tail part. Other groups, such as crabs and lobsters, hatch at a more developed stage (Fig. 21(L)), already having a segmented tail and limbs for swimming. These larvae have long spines on the carapace to help in flotation.

Whatever their form the larvae all drift off in the plankton to spend a longer or shorter time there before settling on the shore or bottom for their adult life. Some eggs are well provided with yolk and the larvae need no other food during their pelagic life. Some are small and the larvae are wholly dependent on the plankton for food; a few are intermediate, having some yolk but not enough to sustain them during the whole of their larval life. It is interesting to consider which type is the most efficient. The question has been discussed by Thorson. The large yolky eggs are, so to say, "expensive" to produce and one female can lay only a few; however, they are not subject to as many hazards as the yolkless eggs and on the whole have a better chance of survival. The small egg is

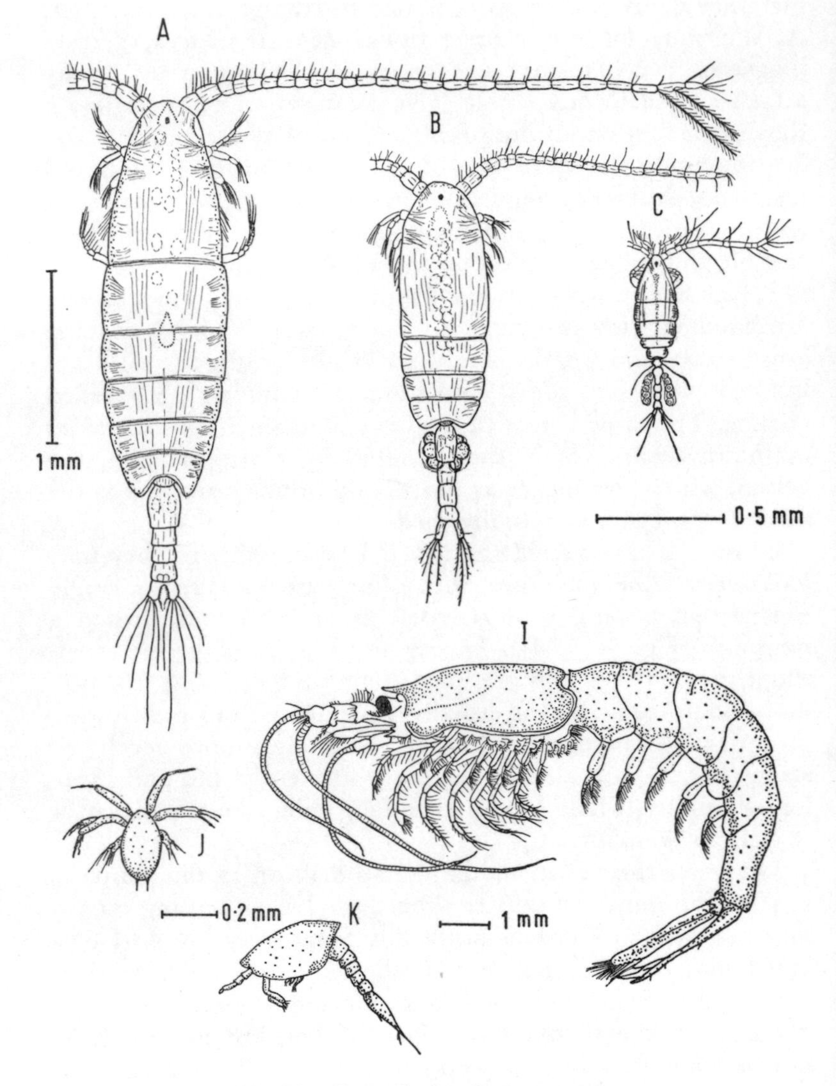

Fig. 21.—Some planktonic crustaceans.

A–F, copepods. A, *Calanus finmarchicus*; B, *Pseudocalanus minutus*; C, *Oithona helgolandica*; D, *Centropages hamatus*; E, *Acartia clausi*; F, *Temora longicornis*; G and H, the cladocerans *Podon intermedius* and *Evadne nordmanni*; I, *Meganyctiphanes norvegica*; J, its nauplius and K, its calyptopis larva; L, zoea larva of shore crab; M, nauplius and N, cypris larva of acorn barnacle; O, amphipod *Hyperia*. All copepods except *Calanus* to the same scale.

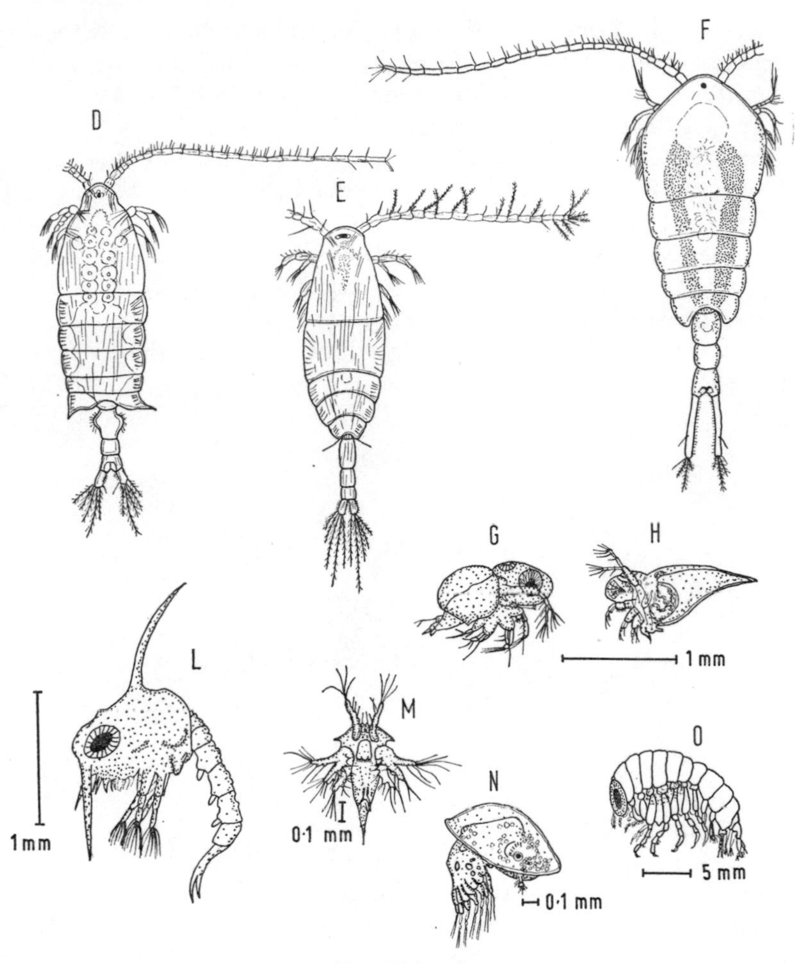

Fig. 21 (*cont.*)

"cheap" and is produced in enormous numbers, but the larvae may not always find enough to eat. Thorson observed that there are much greater fluctuations in the numbers settling from year to year among the larvae derived from the small eggs; in a good year, that is when the food supply is adequate at the right time, there is a very large settlement on the shore, but in a poor year few settle. Settlement of larvae from the yolky type of egg remains much more constant from year to year. Round our shores the small larva feeding in, and dependent on, the plankton is the more common. The time spent in the plankton varies from a few hours to a few weeks.

It is extraordinary how in some years a normally scarce form can become abundant. The Actinotrocha larva of *Phoronis* (Fig. 20(E)) occurs every year in the Clyde sea area; it is never common but specimens turn up occasionally throughout the summer. In 1955, however, every tow-netting taken during the summer contained dozens, if not hundreds, in all stages of development. When they were put in a dish with a little sand the fully developed larvae would settle down, turn themselves inside out, and become tiny U-shaped worms in the course of a few hours (Fig. 22). The adult *Phoronis* is rarely

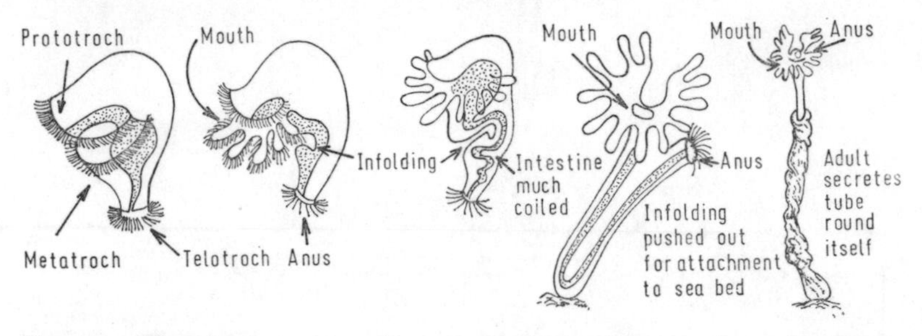

Fig. 22.—The development and metamorphosis of an Actinotrocha larva. (*From Hardy*)

found on Clyde shores and even after this *annus mirabilis* it seemed no commoner than before.

Something must be said here of the Hydromedusae (Fig. 23(A–C)), which spend their sexual life in the plankton and their asexual on the shore or sea bottom. The hydroids, or sea-firs, reproduce by budding off small medusae from special polyps and these swim off and grow in the plankton, eventually producing eggs and sperm. The resulting larva settles and becomes a hydroid again. The larger Scyphozoa (to which group the brown stinging jelly-fish belongs) also spend part of their lives as polyps in shallow shore areas.

Most fish, whether they live as adults on the bottom or in midwater, spend the earliest weeks of their life in the plankton, first as eggs and then as feebly swimming larvae.

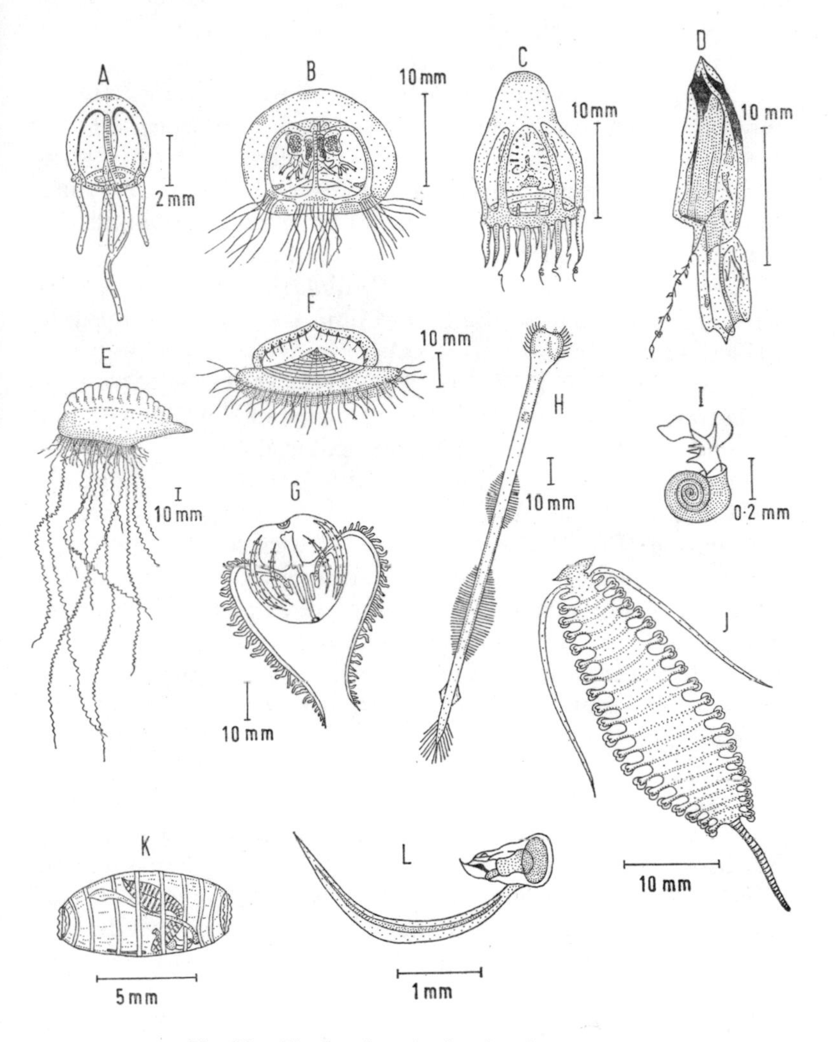

Fig. 23.—Planktonic animals of various groups.

A–C, Hydromedusae; A, *Sarsia tubulosa*; B, *Bougainvillia principis*; C, *Leuckartiara octona*; D–F, Siphonophora; D, *Chelophyes appendiculata*; E, *Physalia physalis*, the Portuguese Man-of-War; F, *Velella velella*, the By-the-wind-sailor; G, a ctenophore (sea gooseberry), *Pleurobrachia pileus*; H, an arrow worm, *Sagitta elegans*; I, a sea butterfly, *Spiratella helicina*; J, a polychaete, *Tomopteris helgolandica*; K, a salp, *Dolioletta gegenbauri*; L, an appendicularian, *Oikopleura dioica*.

We now come to the animals which are planktonic throughout the whole of their lives. Most groups of marine animals have one or more planktonic representatives and one or two in each group will be mentioned.

It is impossible to draw a line between the plant and animal unicellular organisms (protophyta and protozoa) of the plankton. Some of the protophyta can feed both by photosynthesis or on dissolved organic matter. Many of the dinoflagellates (mentioned in Chapter 2) have no chlorophyll and feed by ingesting smaller flagellates or diatoms.

The Foraminifera and the Radiolaria, both one-celled forms making skeletons, the first of lime and the second usually of silica, are found in small numbers in our seas although they are more abundant in the tropics. Their skeletons, sinking to the bottom and accumulating over aeons of time, form layers of deposit on the bottom which are called "oozes".

One animal group which is widespread throughout the world is that of the planktonic ciliates. Among these are the Tintinnidae (Fig. 24), which secrete cases for themselves of very various shapes. The greatest variety is found in the tropics, but a few forms occur in our waters and in summer they are sometimes found in enormous numbers. The animal lives attached by a stalk to the foot of its cup and it moves by a ring of cilia at the mouth. It can abandon its case easily and presumably secrete a new one. Some cases are finely sculptured, some reticulated, some are built with lumps of secretion which look like pebbles and may incorporate extraneous objects such as coccoliths; some are transparent and apparently structureless. Most of the cases are closed at the foot but in some the animal lives in an open tube and is attached to the side. Tintinnids feed partly on dinoflagellates and silicoflagellates.

The Coelenterates have several striking planktonic members. Among these are the Siphonophores (Fig. 23(D–F)), which are really floating colonies made up of several kinds of polyp which have become specialized for different functions. Some are for catching and digesting prey, some are pulsating bells which move the colony along, others are reproductive. In some siphonophores there is a gas-filled float which keeps the colony near the surface; among these is the well known and dreaded Portuguese Man-of-War (Fig. 23(E)) whose bright

Every acceptable catch which gladdens the heart of the fisherman represents the end product of a chain of life—fertility based on sunlight, the nutritive elements in the ocean, and the minute plant and animal life thereby originated and maintained.

That minute life is specifically illustrated in the following pages. Their purpose is to convey the wonder and charm of that life and to emphasize the importance of this chain in providing that fertility which, spreading upward through progressive chains, produces in the end that harvest which inspires and maintains the whole commercial life of the fishing industries throughout the world.

To both fishermen and biologists these pages are of intense interest.

[*To face page* 64.

Plate Ia.—Living phytoplankton. Chains of several species of *Chaetoceros* (those with spines), a chain of *Thalassiosira* (bottom left-hand corner) and *Lauderia* (to the right of the latter). (*D. P. Wilson*)

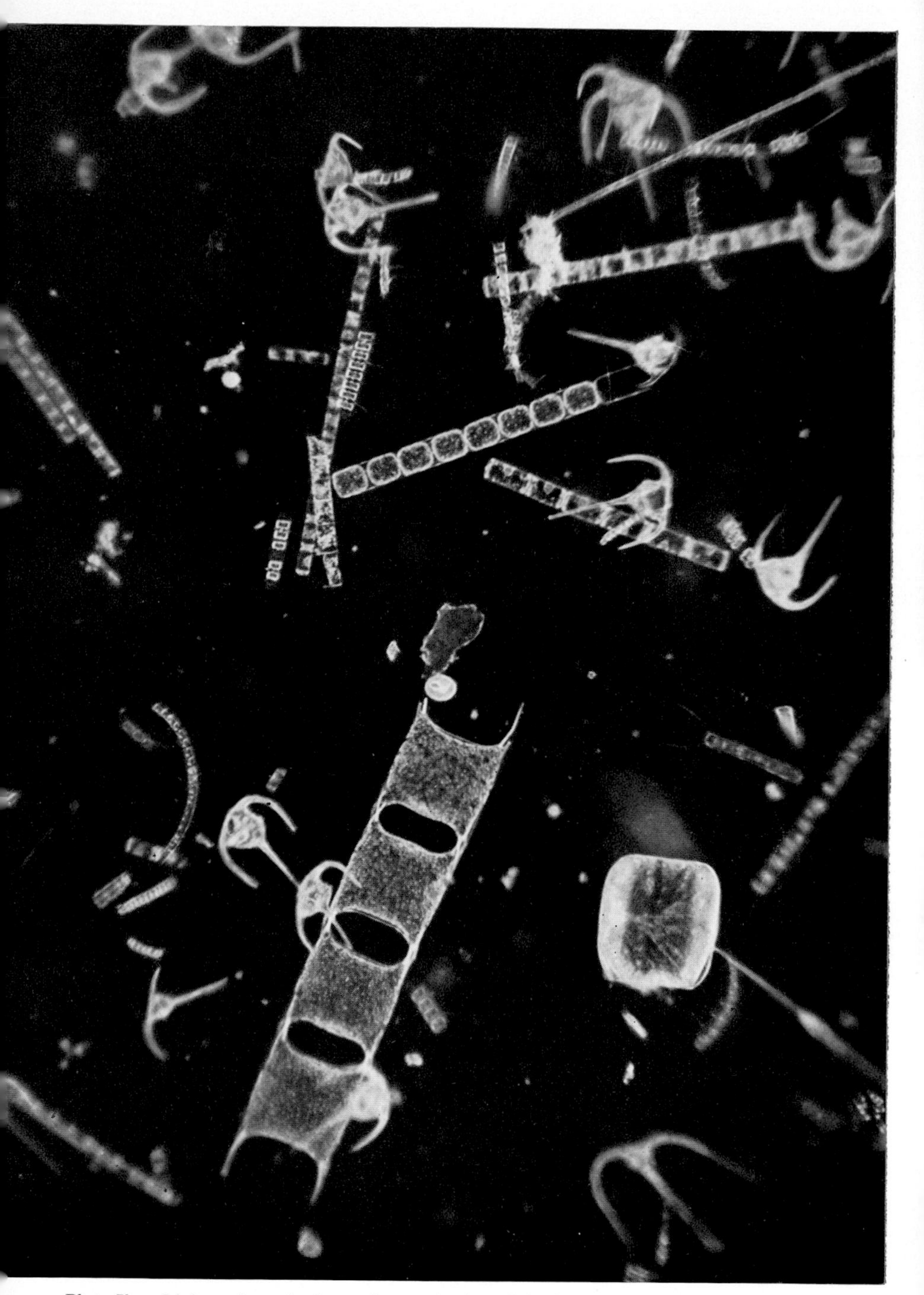

Plate Ib.—Living phytoplankton. Large single cell is *Coscinodiscus*; four linked cells, *Biddulphia*; anchor-shaped cells, *Ceratium*, a dinoflagellate. (*D. P. Wilson*)

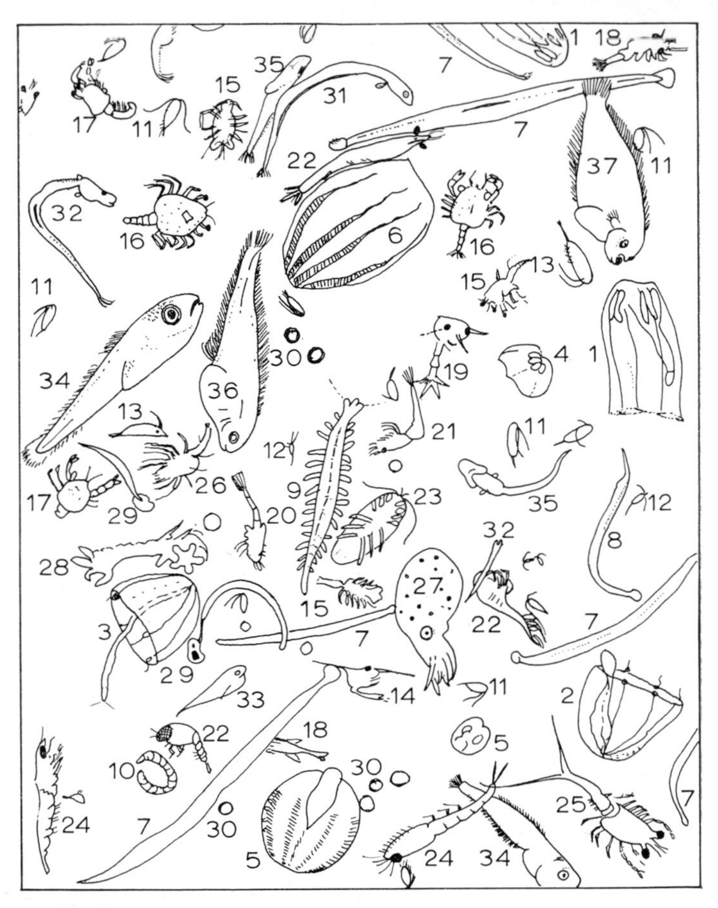

Key to organisms:

1. *Aglantha digitale* (medusa)
2. *Sarsia tubulosa* (medusa)
3. *Dipurena ophiogaster* (medusa)
4. *Bougainvillia* (medusa)
5. *Pleurobrachia pileus* (ctenophore)
6. *Beroe cucumis* (ctenophore)
7. *Sagitta elegans* (chaetognath)
8. *Sagitta setosa* (chaetognath)
9. *Tomopteris septentrionalis* (polychaete)
10. *Poecilochaetus serpens* (polychaete)
11. *Calanus finmarchicus* (copepod)
12. *Metridia lucens* (copepod)
13. *Anomalocera patersoni* (copepod)
14. Zoea of *Corystes* (crab)
15. Megalopa of *Hyas* (crab)
16. Megalopa of *Portunus* (crab)
17. Larva of *Eupagurus* (hermit crab)
18. Larva of *Galathea* (squat lobster)
19. Larva of *Munida* (squat lobster)
20. Larva of *Pandalus* (prawn)
21. Larva of *Nematocarcinus* (shrimp)
22. *Themisto gracilipes* (amphipod)
23. *Eurydice spinigera* (isopod)
24. *Thysanoessa inermis* (euphausid)
25. Larva of *Nephrops* (Norway lobster)
26. Larva of *Sergestes* (oceanic prawn)
27. Larva of *Eledone* (octopus)
28. Metamorphosing *Asterias* (starfish)
29. *Oikopleura* (tunicate)
30. Fish eggs (various)
31. Young herring
32. Young sandeel
33. Young shore sucker (*Liparis*)
34. Young haddock
35. Young cod
36. Young lemon sole
37. Young plaice

Plate II.—Zooplankton haul. A key to help in identifying the animals is given opposite. (*James Fraser*)

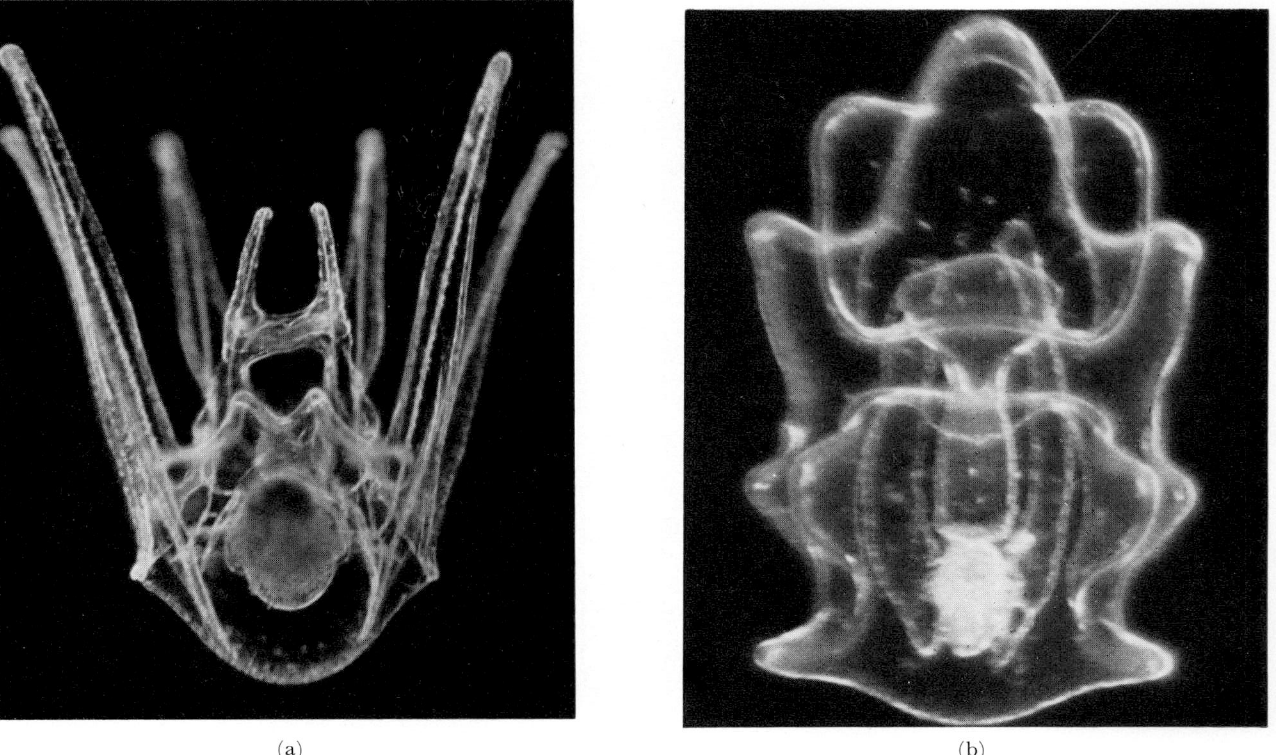

(a) (b)

Plate III—Zooplankton larvae. (a) Echinopluteus; (b) Bipinnaria. (*D. P. Wilson*)

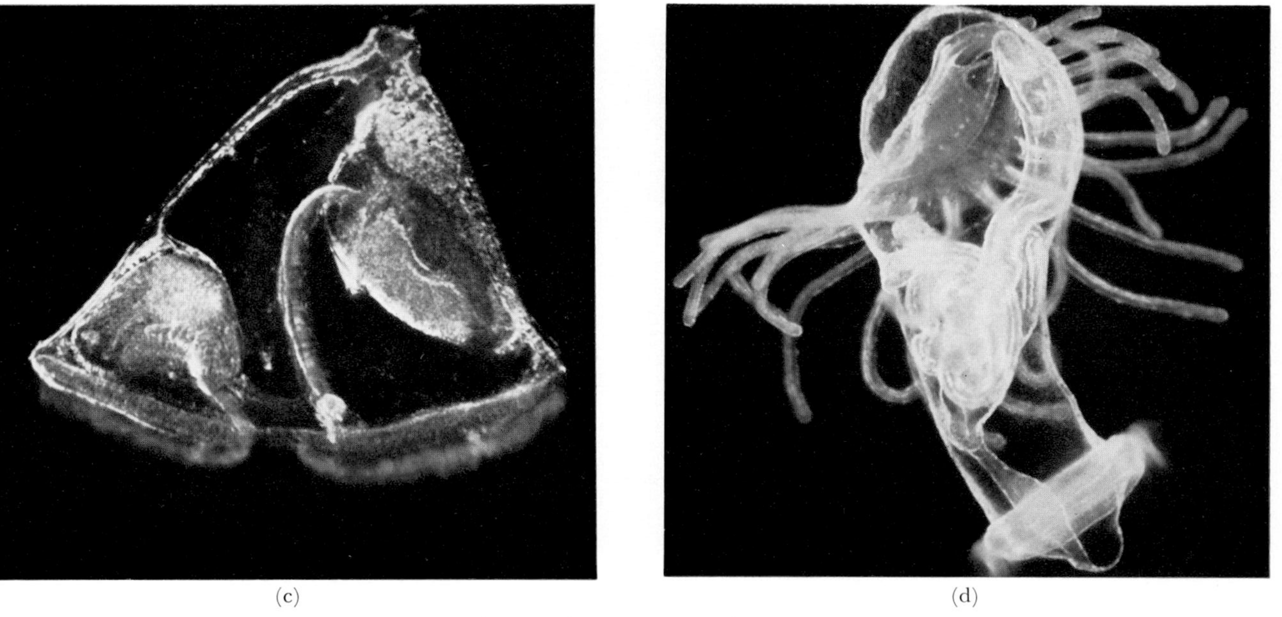

(c) (d)

Plate III.—Zooplankton larvae. (c) Cyphonautes; (d) Actinotrocha. (*D. P. Wilson*)

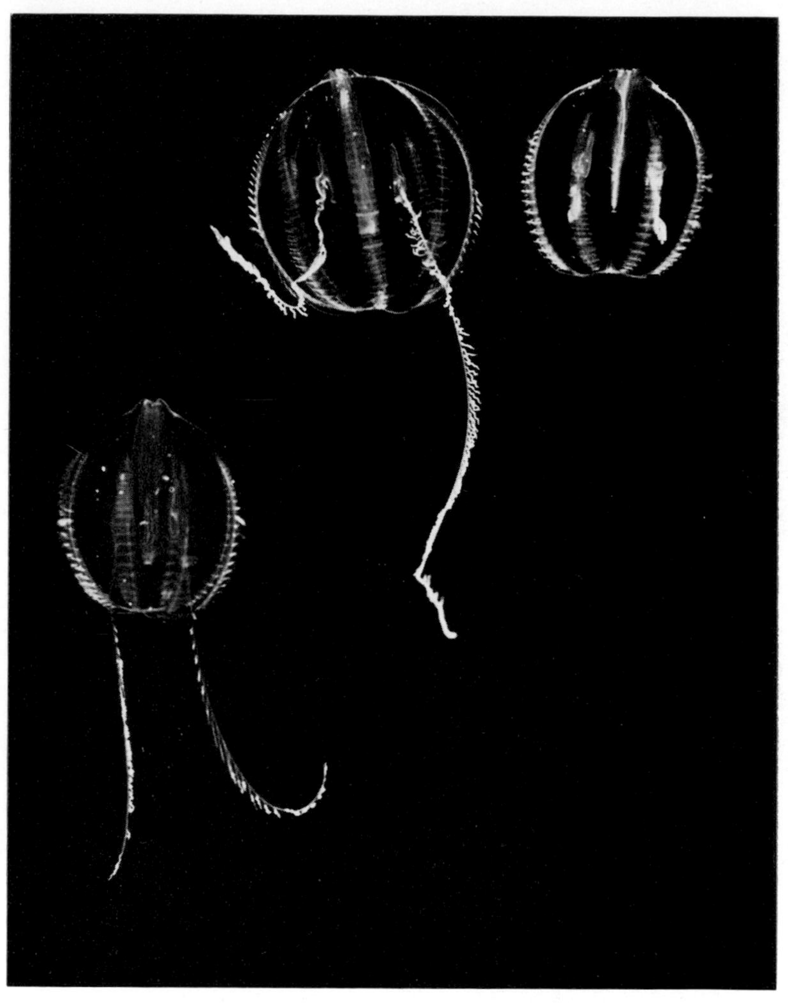

Plate IV.—*Pleurobrachia*. (*D. P. Wilson*)

Plate V.—*Calanus finmarchicus*. (*D. P. Wilson*)

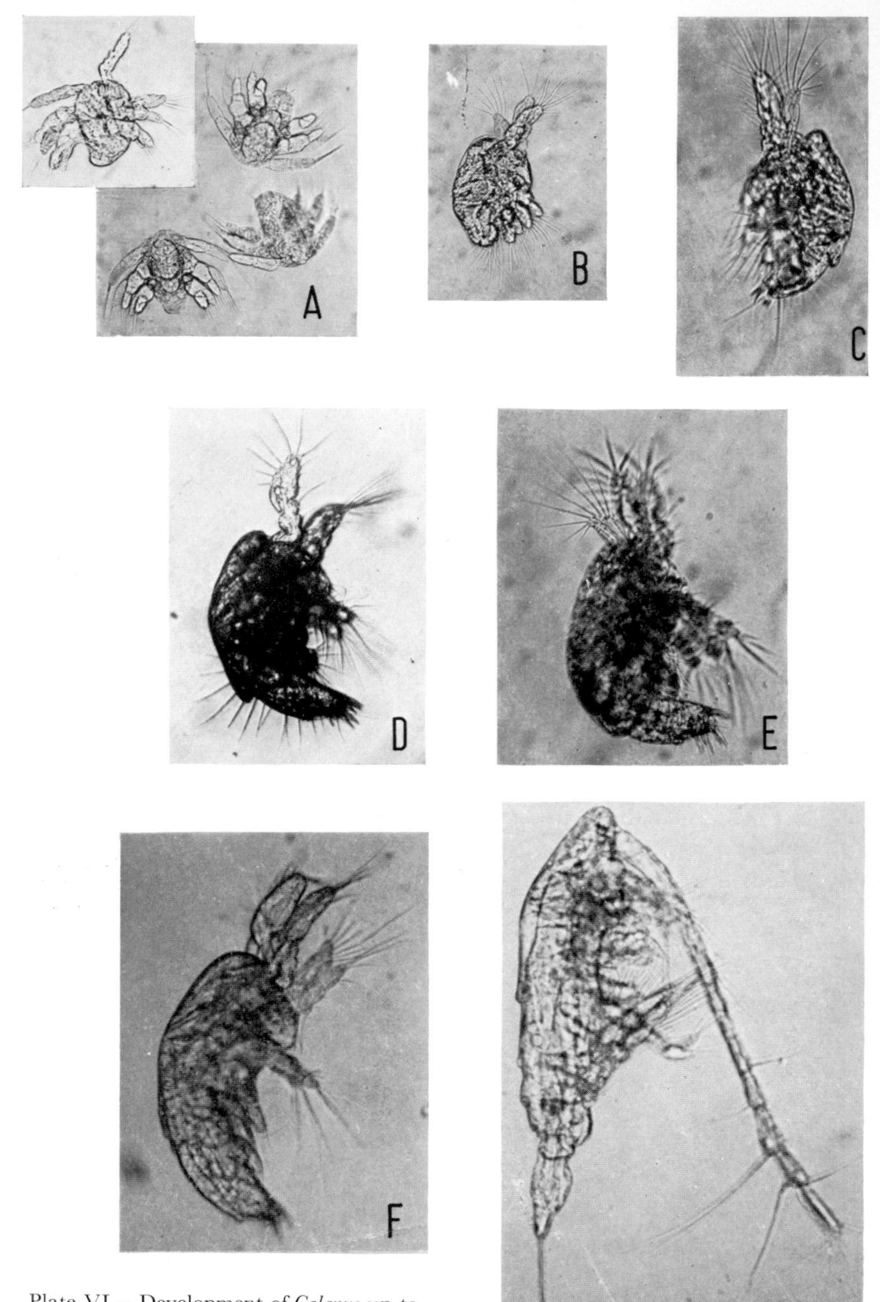

Plate VI.—Development of *Calanus* up to copepodite stage I.

A–F, Nauplius I–VI; G, copepodite I.

Plate VII.—Plankton wheel. (*Hardy and Bainbridge*)

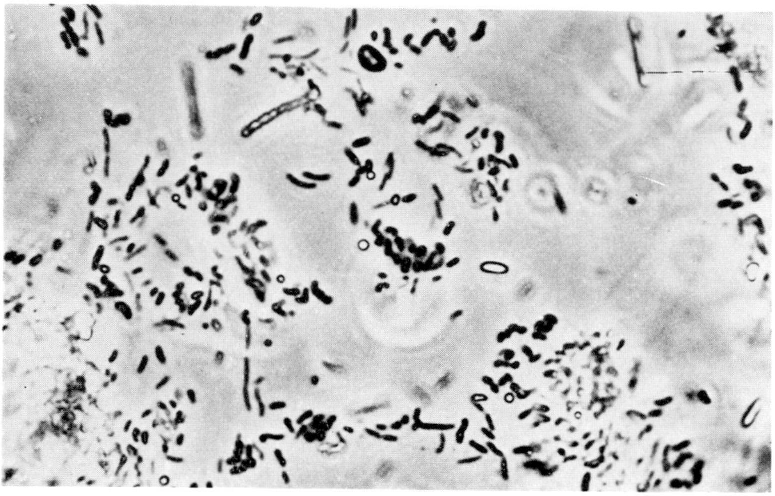

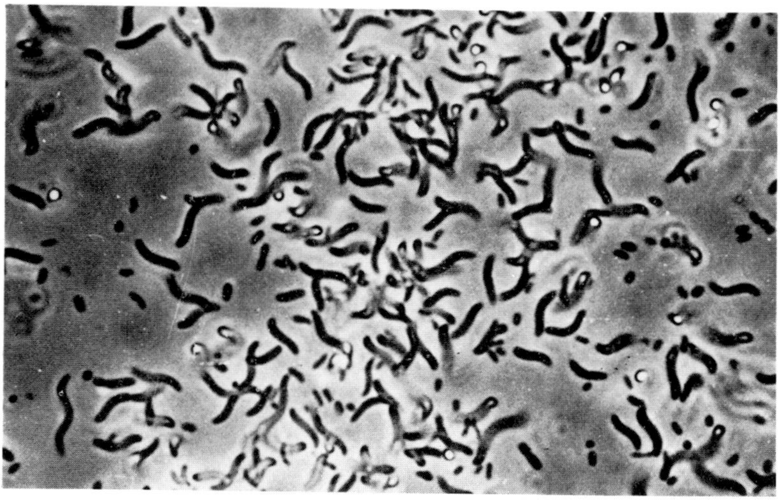

Plate VIII.—Bacteria from decaying marine detritus. (*M. R. Droop*)
Above: various bacteria; *below: Spirillum*, a large highly motile organism.

Plate IX.—Oyster culture. (*C. M. Yonge*)

Above: Oyster parks exposed at Arcachon, showing surface, bare on right, but covered with oysters on left; in background stranded boat, and palisade to keep out rays; *below:* Collector at Arcachon with cultivator inspecting tiles.

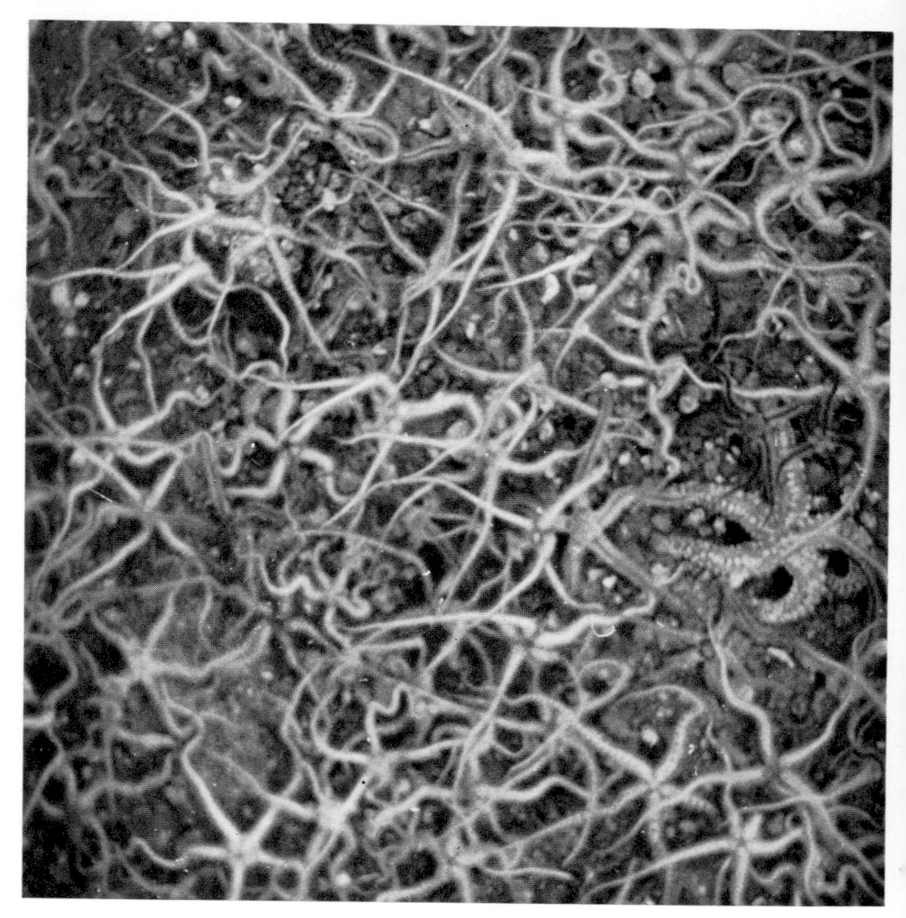

Plate X.—Population of brittle stars on a hard bottom near the Eddystone Lighthouse (about 340 per sq. m). (*H. G. Vevers*)

Plate XI.—Loch Craiglin. (*S. M. Marshall*)

Plate XII.—Fish ponds in Java. Small shaded pond in foreground of lower picture is for fry. *Above* (*C. F. Hickling*) *Below* (*W. H. Schuster*)

blue bladder, triangular and sticking up above the surface, acts as a sail. The tentacle-polyps, extending to a length of many feet, have stinging cells which are deadly to fish and dangerous to human beings too. *Velella*, the By-the-wind-sailor (Fig. 23(F)), with a triangular purple "sail", also floats at the surface and is sometimes blown on to our shores.

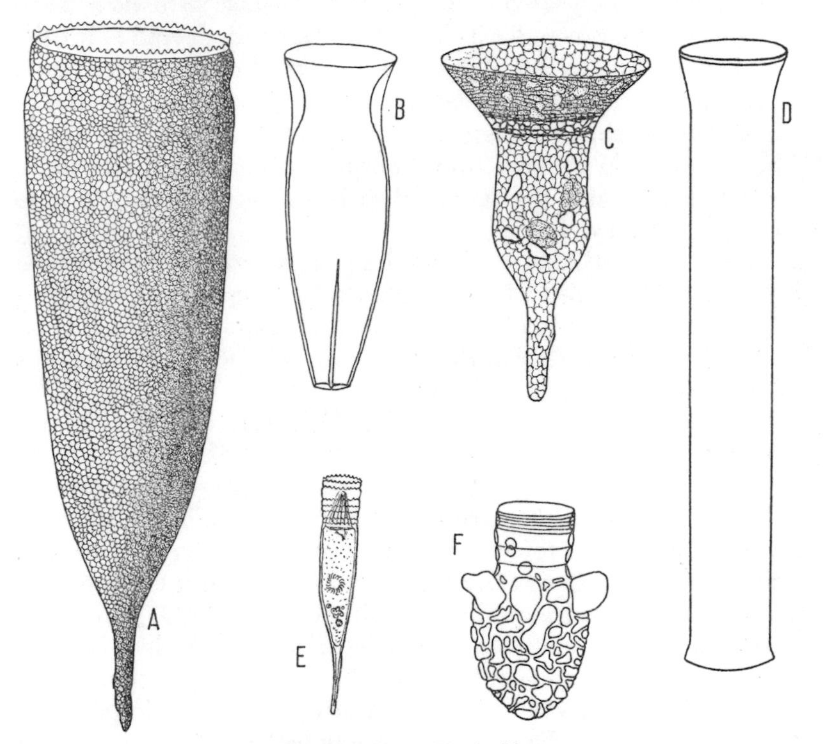

Fig. 24.—Some Tintinnids.

A, *Favella serrata*; B, *Amphorella quadrilineata*; C, *Tintinnopsis campanula*; D, *Tintinnus lusus undae*; E, *Helicostomella subulata*; F, *Codonellopsis indica.*

Another group, much more often seen in temperate seas, is that of the Ctenophores or sea gooseberries (Fig. 23(G) and Plate IV). Made of clear and transparent jelly, they move, not by pulsations of a bell, but by the beating of vertical rows of tiny plates made of fused cilia. The plates have an iridescent sheen and the ctenophore is crystal clear; but lovely as these little creatures are they are most destructive, killing far more than they eat. They have no stinging cells and catch their prey with a sticky secretion. In early summer, when they become abundant, coarse netting has to be tied over the mouth of a tow-net to obtain a catch of unspoiled living plankton.

Of worm-like creatures in the plankton two are outstanding. One is a polychaete *Tomopteris* (Fig. 23(J) and Plate II); it is completely transparent but has the normal form of a polychaete worm, except for its long whip-like tail, devoid of bristles and often broken off. Like so many of the transparent zooplankton creatures, it is carnivorous and will eat crustaceans and young fish. The other, the arrow-worm *Sagitta* (Fig. 23(H) and Plate II), is like a thin transparent rod but it can move very quickly and the powerful jaws in its head enable it to capture its prey.

But the most important members of the zooplankton belong to different families of the crustacea, among them Cladocera, Copepoda, Amphipoda, and Euphausiacea. The most widespread and numerous of these are the Copepoda and they will be treated in more detail later.

Although the Cladocera (Fig. 21(G,H)) are best known from fresh water there are a number of marine species. Since they produce resting-eggs which sink down to the bottom during the winter they are not purely planktonic but they are regularly found in tow-nettings during the summer months and, since they reproduce actively, spread far out into the ocean.

The large-eyed hyperiid amphipods (Fig. 21(O)) are more common in the plankton of northern seas. They live in deep water in the Norwegian Sea and are a food for whales.

The Euphausiacea is the crustacean class next in importance to the Copepoda. Known collectively as "krill" (a Norwegian word given to whale food) they are shrimp-like creatures which may be up to $2\frac{1}{2}$ inches long. They live in deep water and so are not often taken near the coast except in deep lochs and fjords. The eggs are shed freely into the sea, hatch as nauplii and go through a long series of moults, becoming gradually more like the adult. *Meganyctiphanes norvegica* (Fig. 21(I)), a species common in Scottish waters, takes a year to grow to maturity, the eggs being laid in March to April. *Euphausia superba*, the main food of whales in the Antarctic, and one of the most abundant of the plankton animals there, takes two years to grow up. The later larvae live in the cold surface waters and are carried gradually northwards and eastwards. When adult they migrate into deep water to spawn and as there is a current of deep water moving south and west, the early larvae are carried back to the edge of the ice, where they come up again into the cold surface water. So during the course of their life these shrimps make a journey up to sub-antarctic waters and back again. They carry light organs at the base of some of their limbs and perhaps recognize one another by the light pattern formed, for they flash most often when they congregate at the breeding season. They are also a rich source of Vitamin A and this curiously enough is concentrated in their eyes. *Euphausia superba* is a herbivore throughout its life and *Meganyctiphanes*, although smaller, is herbivorous when young but changes over to eating copepods when it becomes adult.

Other crustaceans—ostracods, isopods, cumaceans and mysids—are sometimes seen in the plankton. But most of these are really bottom-living, and come up into surface waters only at night or when they are swept up off the bottom by stormy weather.

The vast majority of molluscs in the plankton are there as larvae only, but there are a few gastropods which have taken to a permanently pelagic life, the pteropods or sea butterflies. Most of them are to be found only in warmer seas than ours, but two genera, *Spiratella* (Fig. 23(I)) and *Clione*, are to be found in temperate waters. The foot, on which most gastropods creep, has been flattened and spread into a pair of wings and it is the flapping of these which gives the "butterflies" their name.

Finally we come to the group of Protochordates, which are related to fishes but have no proper backbone. Salps (Fig. 23(K)) are the planktonic cousins of the sea squirts on the shore; they live within a gelatinous or leathery cylindrical envelope and often one is joined to another so that long strings are found drifting about. When salps (or sea squirts) reproduce, their eggs hatch into larvae rather like tadpoles with a big head, a stiffish rod to act as backbone and an active tail. They also have a primitive eye. Thus the larva shows its relationship to vertebrates more clearly than the adult. The second kind of protochordate found in the plankton, the Appendicularian (Fig. 23(L) and Plate II), is a larva which has become adult without changing its larval form. It has however secreted a gelatinous house (not shown in the figure), one part of which is constructed as a very fine filter. Through this, by means of vigorous flapping of its tail, it draws a current of water and screens off the smallest plankton organisms as food. It was from these filters that some of the minute flagellates and protozoa were first described.

Zooplankton have problems of flotation, for although they usually have some means of locomotion it is rather feeble. Some of the adaptations found in phytoplankton are found in zooplankton too, for instance the great increase of surface by the production of bristles. Some warm-water copepods have very long feathery caudal furcae. Some of the zooplankton use chemical means. Replacing the sodium sulphate of the body fluids by sodium chloride reduces the specific gravity and makes the animal more buoyant. This is found particularly in gelatinous animals such as jellyfish, ctenophores, salps and pteropods.

We return now to that most important group of Crustacea, the copepods. They are important because they are so numerous and so widely distributed and act as a link between the phytoplankton and the fish, turning the phytoplankton into a food palatable to these larger carnivorous creatures. They are preyed on not only by fish and whales but by many larger plankton animals which in turn make food for fish. There is nothing which quite corresponds to them on land. There, most of the herbivores are large and can be eaten directly by man; copepods, although a dish of them fried tastes like shrimp paste, are too small to be conveniently collected and so we leave it to the fish to concentrate them for us. Copepods are not planktonic only; they are ubiquitous and are found in sand, in rock pools, among seaweed, and as parasites in or on fish and invertebrates.

Of the planktonic forms there are hundreds of species scattered over all the oceans of the world, but in any one place there will be perhaps half a dozen species which occur constantly and are abundant in their season (Fig. 21(A–F)). One of the most widely distributed and abundant genera is *Calanus*

(Plate V); it is also large for a copepod, being 2–5 mm long, and for this reason most of the research work done on copepods has been concentrated on it. However, a great deal of what has been found out about *Calanus* will be true of many other planktonic copepods too.

The body is divided into a more or less cylindrical front part, the metasome, and a segmented narrow tail, the urosome. The front end has a simple median eye and bears a pair of long antennae; there are then five pairs of jointed limbs, grouped round and behind the mouth, and behind them, in series on the metasome, five pairs of swimming feet. The urosome usually has no limbs and ends in a forked process with long feathery bristles on it. Male and female differ slightly from one another in appearance, mainly in the feathering of the limbs.

When *Calanus* is still, or moving gently in the water, it supports itself and swims by the movement of the limbs round the mouth. When it is touched and wants to escape—and it can make surprisingly great leaps and capers—the swimming legs, normally folded up ventrally, jerk outwards, the antennae beat down as well and the urosome gives a flick.

By the steady beating of the mouthparts a current of water is drawn from behind forwards and passes through a pair of limbs, the maxillae, which are finely feathered and filter off the particles in the water. Other limbs scrape the maxillae and push the food to the mouth. The food is then churned up in the gut by vigorous peristalsis and, after the nourishment has been absorbed, the remains are compacted into a firm oval body at the hind end of the gut and ejected as a faecal pellet.

The process of filter feeding goes on more or less automatically all the time, but the copepod need not necessarily swallow all the food collected. Although they will take in particles of Indian ink or other indifferent matter, yet there are hours when nothing enters the gut at all although the copepod has food available and seems to be swimming about quite normally. Perhaps it can scatter the collected food again with a flick of the mouth parts. Filtering is not the only method of feeding. *Calanus* and other copepods have been seen catching organisms, or seizing a large diatom and sucking out the contents. That filtering is, however, the normal method of feeding is suggested by several facts. First, if a *Calanus* is kept in an increasingly rich concentration of diatoms or flagellates it produces more and more faecal pellets, showing that it goes on filtering about the same quantity of water whether it is in a rich or a poor medium. The same thing is shown in the sea during a diatom increase, when copepod faecal pellets increase according to the number of diatoms and not according to the number of copepods. Secondly, when the gut contents of *Calanus* were examined throughout a whole year, the food it had eaten was found to agree pretty closely with the microplankton in the sea at the time it was caught. However, when food is scarce there is no doubt that a copepod has other means of getting it.

A copepod hatches from the egg as a nauplius and like most crustacea goes through a series of moults and stages before it reaches adult form. With *Calanus* there are six nauplius stages (Fig. 25), and six copepodite, the last of which is the adult. At each stage it acquires more limbs or more bristles on the limbs already present. During the winter in temperate latitudes the *Calanus* population lives in deep water and is almost all copepodite stage V, that is, the stage immediately preceding the adult. In January moulting begins; males appear first and then females. Females are fertilized by the attachment to them of a spermatophore, a sausage-shaped bag containing sperm, soon after they have moulted, but they are not yet ready to lay eggs and the sperm are stored in two sacs, the spermathecae, near the opening of the oviduct. When phytoplankton is scarce, as in winter, it takes several weeks before the eggs are ripe, the time depending very much on the food available. In summer, when the stage V have been feeding freely, a female may take only a day or two after moulting before she lays eggs. In spring most females are ready to lay by March. Although a few eggs may be laid and develop before then, the main spawning takes place in March which is usually the time of the spring diatom maximum. The spawning of copepods often coincides with diatom increases at other times of the year also. This is understandable because the number of eggs laid by an individual *Calanus* female depends very largely on the amount of food she has eaten.

A *Calanus* will lay 20 to 80 eggs at a time and if well fed will continue to lay up to a total of several hundred. The eggs are

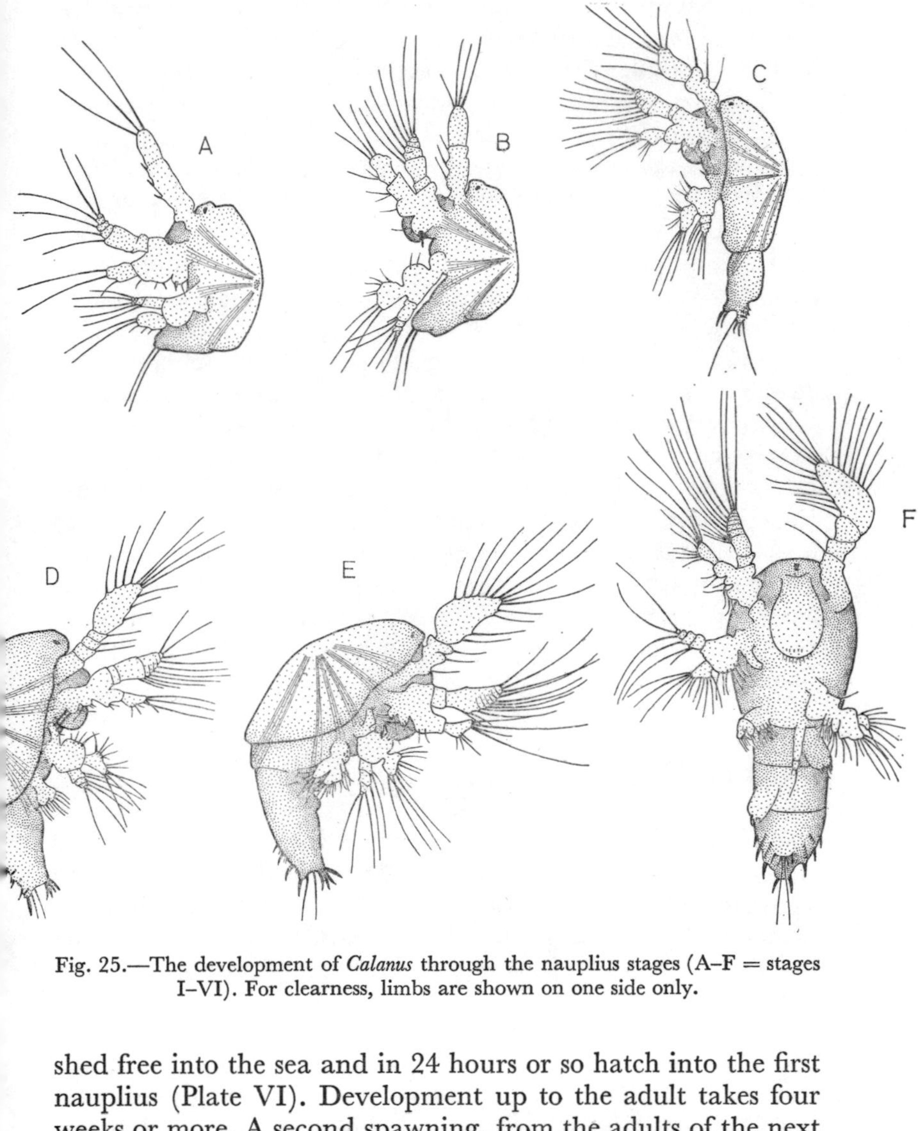

Fig. 25.—The development of *Calanus* through the nauplius stages (A–F = stages I–VI). For clearness, limbs are shown on one side only.

shed free into the sea and in 24 hours or so hatch into the first nauplius (Plate VI). Development up to the adult takes four weeks or more. A second spawning, from the adults of the next generation, takes place about May, and others may follow later in the summer. In the autumn, however, the developing copepodites do not moult into adults but accumulate as stage V to form the over-wintering population,

This pattern of development is true of the waters round our coasts but in other latitudes it may be very different. In a fjord in East Greenland where *Calanus* was studied throughout a year, there was only one spawning during the year. It happened in July along with the annual diatom increase. In these cold arctic waters development goes on much more slowly and by autumn the population was made up of copepodite stages III, IV, and V and it then moved down to deep water for the winter. There, in December and January, those which had reached stage V moulted into adults and the females were fertilized. When the population moved to the surface again in May or June the younger stages had still to complete their development, which they did during the summer. Spawning was therefore spread out over the brief summer and it is possible that some of the youngest copepodites did not have time to become adult, and so spent a second winter still immature. However, in the main the species is annual. It is annual also in the Barents Sea. Off the north-west coast of Norway there is one spawning in spring and possibly another in the summer. In other places the pattern of the life cycle can be fitted in between these two extremes of 2–3 months and more than a year.

Several interesting points come out of a study of these facts. First, the moult of the over-wintering generation into adult comes at the same time of year in both arctic and temperate water; it is therefore not caused by temperature. Then, it is possible for there to be a long delay between fertilization and egg-laying, six months in Greenland *Calanus*. Finally the length of life of an individual *Calanus* is very different in different climates and even in different seasons in the same area, from two or three months in summer in temperate regions to more than a year in arctic seas.

The different generations of *Calanus* (and this is true of plankton copepods as a whole) are not all the same size. Those born and brought up in cold water are larger than those born and brought up in warm water. Even in the same region the size varies from summer to winter. It may seem paradoxical that the over-wintering generation is the smallest, but then they were born during the warmest summer months. The eggs laid in March, when the sea is at its coldest, grow up into the biggest *Calanus* of the year. An average total length for a female *Calanus* in the Clyde sea area is 3·1 mm (summer) to 3·7 mm (spring); in Greenland waters it is from 3·2–5·4 mm. There are other species of *Calanus*, both larger and smaller.

A careful study of the size of copepods in different regions showed that, when temperature remained fairly constant and the amount of food varied, the copepod size depended on the food. When food was always sufficient and temperature varied, the size depended on the temperature but changed in the opposite direction. When both food and temperature were variable, both had an effect on size.

It is a curious fact that, in a homogeneous population of *Calanus* of one stage, measured at one spot, the largest animals are found in the deepest water. Considering that the whole population has been hatched at the same temperature, and that they may migrate daily through the different temperature layers, it is astonishing that they manage to keep themselves sorted out in this way. Euphausiids do the same, although with them the question of age may be a factor.

The position of rest of a copepod in the water is hanging vertically with its head uppermost and its antennae extended. Most of its movements are therefore in a vertical direction rather than a horizontal, and the migrations they undertake are to and from the surface.

These migrations may be seasonal and many copepods spend part of the year in deep water. *Calanus* spends the winter there, coming nearer the surface in spring. "Deep" and "surface" water may mean different things in different places. Observations made from the weathership in the Norwegian Sea have shown that *Calanus* may go down below 1000 m in winter; off the Lofoten Islands the winter depth is below 200 m; in the Norwegian fjords and the Clyde sea area it is below 100 m. After the spring spawning the depth distribution is less constant. In the Clyde sea area the first generation in May, and in the English Channel a later generation in July, are found right at the surface.

Other animals show seasonal migrations too, for example the antarctic *Euphausia* already mentioned. The polychaete *Tomopteris* in the English Channel lives very near the bottom except in July and August, so that it may seem to be rare or absent at times when it is really quite abundant, but below the reach of the net.

Apart from these seasonal changes many zooplankton animals migrate daily from a daytime depth, which is usually below the photosynthetic zone, to the surface at night. They come upwards as daylight declines, spending the dark hours congregated near the surface, or randomly distributed. At dawn they move downwards again, sometimes (if they have been randomly distributed) rising a little before doing so. This is what might be called an ideal picture but it is not found very regularly. Not all developmental stages behave in the same way and often a species behaves differently in different seasons. Some migrating animals do not reach the surface. Among all groups of zooplankton, including larvae, some species migrate and some do not; it has been suggested that the latter cannot swim strongly enough.

Of those species that migrate, some normally live fairly near the surface and have not far to go; some live in deep water and must swim fast and far to reach it. The speed of a number of species has been measured in the plankton wheel (p. 77). Considerably higher speeds were recorded over a two-minute period than over half an hour, showing that these animals are capable of bursts of speed although they do not maintain them for long. As was to be expected the speed of swimming down was greater than the speed of swimming up. Most of the animals whose speeds were measured were crustacean. The euphausiid *Meganyctiphanes* was the fastest: over two minutes swimming upwards it attained the rate of 173 metres/hour, and over 60 minutes 93 m/h; going down the rates were 215 and 129. The copepod *Calanus* can swim upwards at 66 m/h and maintain an average of 15 m/h for an hour; going downwards comparable rates were 107 and 47; even the nauplius can swim at a rate of 44 m/h. For a small copepod (*Acartia clausi*) the upward rates were: 34 and 9 m/h. Such figures enable one to understand the mechanics of the migrations which go on; it is usually the strongest swimmers which live deepest.

To repeat a series of vertical hauls 6 or 8 times within 24 hours and afterwards to count the zooplankton (or even one species) in them all so as to plot the depth distribution is a time-consuming work, and this is probably why it has been

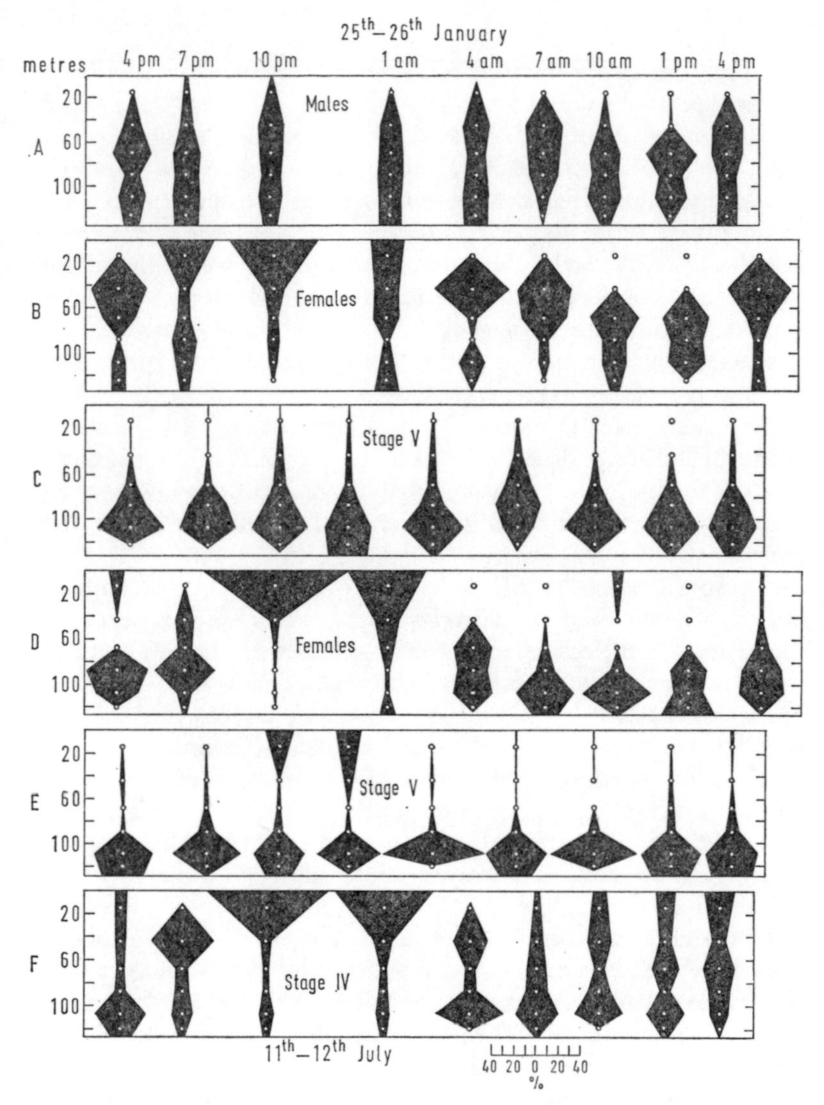

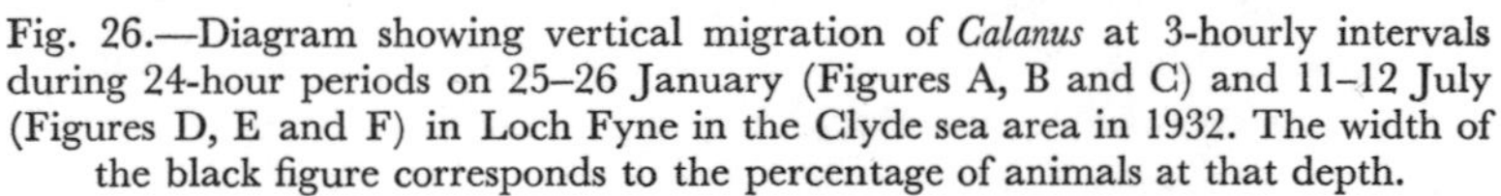
Fig. 26.—Diagram showing vertical migration of *Calanus* at 3-hourly intervals during 24-hour periods on 25–26 January (Figures A, B and C) and 11–12 July (Figures D, E and F) in Loch Fyne in the Clyde sea area in 1932. The width of the black figure corresponds to the percentage of animals at that depth.

	Sunrise	*Sunset*
25 January		4.27 p.m.
26 ,,	8.37 a.m.	
11 July		8.27 p.m.
12 ,,	3.58 a.m.	

(*After Nicholls*)

done comparatively seldom, and never repeatedly throughout the course of a year. The fullest information has been gained for a number of species in the English Channel, the Clyde sea area, the Cape Cod area and round Bermuda. As an example of the variation which occurs even within one species we shall again take *Calanus*. Russell made a large number of hauls in the English Channel at different times of the year, including several 24-hour series, and found that the adults were usually deeper in sunny than in dull weather and that their diurnal movements approximated to the "ideal" pattern. In July males lived nearer the surface and at a higher light intensity than during the rest of the year. Nicholls made two 24-hour series (Fig. 26) in the Clyde sea area, one in July and one in January, thus sampling what must have been the over-wintering generation at the beginning and end of its life. In winter the females showed the greatest migration, males showed a little and stage V none at all. In summer there were not enough males to count, and the females migrated as before, although they stayed a little deeper in the water during the day than in winter; a few stage V came to the surface in the dark and so did a few stage IV, which were otherwise fairly evenly distributed in the water. Earlier stages remained near the surface all the time. However, the *Calanus* of the first generation, which become adult in May, are often found right at the surface even in bright sunshine, and this phenomenon is quite often seen in Norwegian waters too. Euphausiids also have been found swarming at the surface.

In the Gulf of Maine Clarke made a series of observations in July, first over two days in deep water, and then, the next day, in shallow water. At the deep station the *Calanus* migrated regularly from below 100 m right up to the surface and back, the young stages along with the adults. At the shallow station one day later *Calanus* did not migrate at all, most of them remaining above 20 m the whole day; but another copepod, *Metridia*, migrated normally.

It can be seen from all these observations that the behaviour of *Calanus* is by no means uniform and this can be concluded also from experiments made by Hardy and Paton. They lowered *Calanus* into the sea in paired glass cylinders; these were divided into upper and lower halves which could be either separate, or united as one long tube. The *Calanus* were put into the upper half of one cylinder and the lower half of the other, with the halves separate. When the cylinders had been lowered to the required depth the two halves were put in communication with one another, and after a certain time they were separated again and hauled up. Counts of the final distribution showed that, in any population of *Calanus*, some animals were always moving up and some moving down. The proportion of the first was greater as the depth of the experiment increased.

This apparatus eventually developed into the "plankton wheel" (Plate VII), a continuous cylinder in the form of a complete circle or tube in which an individual animal can be studied by turning the wheel so as to keep it always under the observer's eye. The movement of the wheel is recorded on a rotating smoked drum (not shown in Plate VII), by means of a needle geared to the wheel. The effect of depth or pressure is eliminated, but light and temperature can be altered experimentally. The rates of swimming quoted above were all measured in this instrument. Observations with it showed that, as usual, different individuals reacted differently. *Calanus* would not always react to an artificial alteration in light intensity, but under normal conditions it rarely swam upwards during daylight hours or downwards during dusk or dark.

One antarctic copepod, *Pleuromamma robusta*, has a well marked diurnal migration. As we have seen (p. 67), the surface waters in the Antarctic are moving northwards while the deep water is moving south; by its alternation between the two layers *Pleuromamma* manages to stay continuously in much the same area.

Many explanations have been put forward for the cause of these vertical movements. It is obviously closely associated with light, and one explanation is that each species and developmental stage has an optimum light intensity within which it tends to stay. This provides a good explanation for the majority of migrations although, as can be seen from the account above, there are many exceptions. In the waters round Bermuda, both temperature and pressure seem to be controlling factors as well. Those animals living at great depths are less sensitive than the shallower-living, for they are not only more diffusely distributed during the day but on their migrations they pass through a much greater range of temperature and pressure differences. Gravity is another factor which must exert some influence. However, gravity is not the sole cause of downward movement, for zooplankton animals do not sink passively but turn and swim actively downwards.

There have been occasions on which a marked thermocline has seemed to act as a barrier to migration, but more often it makes no difference.

Vertical migration is so striking a phenomenon, and is so widespread among pelagic animals, that one would expect it to have some marked advantage but there is no consensus of opinion on what this may be. Since the zooplankton feed mainly on the phytoplankton they must come to the surface to do so, and presumably it is better to do so at night when they themselves will be less visible to predators. Strong light has a lethal effect on many animals and this may also be a reason for avoiding the surface layers during the day. To be able to move up and down in the water with regularity an animal must have some guide to the depth at which it is at any time. Few animals have been found to react to pressure, and a sensitivity to gravity would have the disadvantage of depending on the weight of the animal—which in turn would vary with its size as well as with the weight of food it had eaten. The simplest guide to depth is light, which decreases rapidly with increase in depth. Harris indeed has suggested that, to maintain themselves at a suitable depth, zooplankton organisms do use light as a guide and that vertical migration is an inescapable consequence of this. The hypothesis has been supported with some very interesting experiments on one of the freshwater Cladocera, *Daphnia*.

Hardy, however, believes that vertical migration has been evolved to give the relatively feeble and slow-moving zooplankton some degree of independence of their environment. In the sea the water at different depths is usually moving at different speeds and often in different directions. By moving into and out of these different layers of water for varying times, an animal is able to sample a far greater variety of conditions than it could otherwise do. If it finds a particularly suitable environment it can stay in it. This Hardy calls "plankton navigation", a name particularly suitable to the migrations already described of *Euphausia superba* or *Pleuromamma robusta*.

More recent hypotheses have tended to stress the advantages of vertical migration for reproduction. David suggests that by means of it populations of a species are (as Hardy has pointed out) broken up and well mixed with others, so that genetic recombination is ensured. Wynne-Edwards supposes that the concentration in the surface layers, especially marked during breeding seasons, enables the copepods "to test the population density and stimulate responses which will hold it at or restore it to the optimum". Both these hypotheses are rather speculative and have as yet little factual evidence in support of them. McLaren's hypothesis is that by spending the non-feeding time in deeper and cooler water an animal will grow larger and because of its increased size lay more eggs and so increase fecundity; but the advantage would appear to be slight.

The puzzles of vertical migration are by no means solved, and much more research is needed before we can understand all its complicated variations and their causes.

Although many zooplankton species have a very wide distribution over the oceans there are some which are limited by temperature, salinity or other factors and are found only in restricted areas. Sometimes the restriction applies to part only of the life-history, such as reproduction; for instance, an animal may be carried by currents into an area where it may survive but cannot breed. The continuation of the species in the new area depends therefore on an annual supply from its original home. Such animals are found, for example, in the Gulf of Maine; the mollusc *Spiratella helicina* (Fig. 23(I)) and the appendicularian *Oikopleura vanhoffeni* come in from the north with the Labrador current but disappear when the Gulf warms up. The copepods *Calanus hyperboreus* and *Metridia longa* are visitors from the Arctic too but may spawn in the Gulf occasionally. There are also visitors from the tropics and the deep water to the south. The arrow-worm *Sagitta serratodentata* and the euphausiid *Thysanoessa gregaria* are examples; they occur in the Gulf most years but do not breed there. In the same way *Physalia* (Fig. 23(E)) or turtles are sometimes carried, with the North Atlantic Drift, from the tropics into the English Channel or even further north up the west coast of Britain.

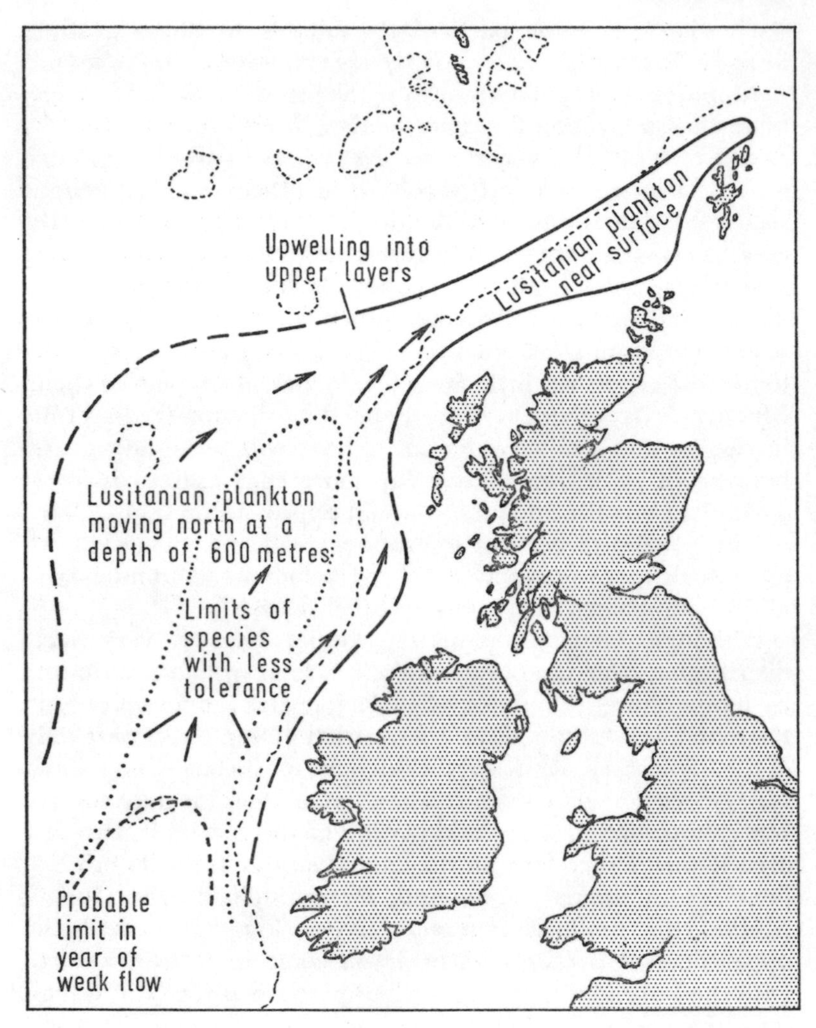

Fig. 27.—Chart showing the northward flow of "Lusitanian" plankton. (*After Fraser and Hardy*)

The fact that many plankton animals are limited in distribution to one particular type of water allows us to use them as labels for that water. Species which are fairly large and easily distinguishable with the naked eye are the most useful, and arrow-worms (*Sagitta* species) are among the best of these "plankton indicators".

Sagitta setosa is a coastal species and *Sagitta elegans* (Fig. 23(H)) is one which, though found close to land, likes some mixture of oceanic water. The first is characteristic of water rather poor in plankton and unsuitable for the rearing of certain larvae in the laboratory. This type of water often fills the English Channel. *Sagitta elegans* is found in the south-west approaches to the Channel, where the water is in general richer in plankton and gives a better environment to developing larvae. The factors which cause the differences between the two types of water are unknown, but each water can be distinguished by the species of *Sagitta* living in it.

The plankton of any area usually contains elements from

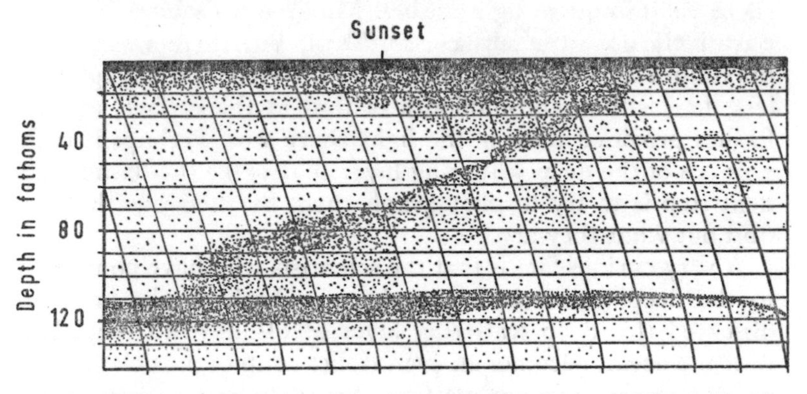

Fig. 28.—Echo-sounder trace of the deep scattering layer off Miami, Florida, showing the evening ascent from the bottom. (*After Moore*)

several sources. The North Sea plankton, for example, contains, besides the indigenous fauna, animals coming round the north of Shetland with the surface waters of the North Atlantic Drift. Below this water there flows a deeper and more variable current coming through the Straits of Gibraltar from the Mediterranean and flowing up the west coasts of Europe and the British Isles (carrying "Lusitanian" plankton, Fig. 27). It eventually rises to the surface and in some years enters the North Sea, again round the north of Shetland. There are also species coming from the colder Norwegian Sea. The water from these different sources is recognized by the "indicator" species found.

No account of the zooplankton would be complete without mentioning that curious problem of the open seas—the deep scattering layer. By using echo-sounders, echoes were obtained from far above the sea bottom which did not look like fish traces. They were found nearly everywhere in deep water except, oddly enough, the Arctic and Antarctic. What puzzled observers was that the layer came up at night and went down by day just as the zooplankton does (Fig. 28). Various animals have been suggested as causing it—euphausiids, small fish, small squids. The interesting point revealed by these observations is that the deep oceans are not so barren of life as had been supposed. This has been shown too by measurement of light at great depths. Using an extremely sensitive photometer, Clarke and his collaborators found that at some depths there were frequent light flashes. Maximum flashing was comparatively near the surface, at 70 m, but there was a lesser maximum at a depth of about 900 m. Here the daylight was less than a thousand-millionth of that at the surface, and the flashes provided most of the light there was. Perhaps it is the animals of the deep scattering layer that make the flashes. The next step is to try to photograph them by their own light, to find out what they are.

A very curious fact is that the scientists who have gone down to the deep waters in bathysphere or bathyscaphe have not observed this deep scattering layer at all. Instead they think that the ocean becomes an even richer soup of particles as they go deeper. Hardy suggests that the particles are perhaps not living but are cast-off moults, and dead plants and animals, sinking slowly from above. Only more observation can explain these contradictions.

Chapter 5

Zooplankton Production

Wherefore did Nature pour her bounties forth
With such a full and unwithdrawing hand,
Covering the earth with odours, fruits and flocks,
Thronging the seas with spawn innumerable,
But all to please, and sate the curious taste?

JOHN MILTON, "Comus"

Most fish are not herbivorous but carnivorous; their link with the phytoplankton therefore goes through the zooplankton. This is true of almost all fish in their young planktonic stages and of adult pelagic fish such as the herring, mackerel or basking shark. Other fish, which take to bottom-feeding in adult life, depend on the phytoplankton more indirectly. It is therefore important to find out about zooplankton feeding: what and how much they need, how much they actually eat and how they use it.

Usually one of the first things we find out about an animal is what it eats, and there is therefore a great deal of scattered information about the food of different zooplankton animals. Some (usually the larger forms, such as jellyfish and some euphausiids) are carnivorous, some are omnivorous, and some herbivorous. To find out what part they play in the productivity of the sea we need however to have more detailed and quantitative knowledge, and this we have for very few species. Because of their importance as food and their abundance we know most about copepods, and among copepods *Calanus* again has been most studied.

A copepod generation lasts for a few months but a phytoplankton increase usually lasts only a week or two, and it is therefore inevitable that copepods have alternate feasts and fasts. Their reproduction is to some extent geared to the diatom increases but there are bound to be times when their food is scarce, most of all during the winter. We therefore want to know how far their requirements are met by what is available in the sea.

A copepod's needs can be roughly estimated from its respiration, for the oxygen it takes up is used to metabolize its food and provide energy. Respiration has been measured in a number of copepods and the results can be expressed in several ways, as the oxygen used per copepod per hour or per day, as the oxygen used per gramme of copepod weight, or as that percentage of the copepod's weight which has to be replaced each day.

One of the first workers on copepod feeding and respiration was August Pütter. He measured the oxygen consumption in batches of mixed copepods, and his estimations were very high. This led him to suppose that, since there was not enough food present in the water as living organisms, copepods, and indeed most marine animals, must be able to absorb dissolved material from the sea water through the surface of their bodies. Later research workers found that Pütter had much over-estimated both the oxygen used up by copepods and the amount of dissolved organic matter in the sea water. It does remain true however that more organic matter is present in solution than in particulate form, and that there are times of the year when phytoplanktonic food seems insufficient for the zooplankton population.

Some very careful work showed that the quantity of dissolved organic matter which could be taken in through the surface of a copepod's body was negligible, but for all that the question of its use in one way or another keeps turning up.

Like the phytoplankton, animals need trace elements and vitamins and, although they normally get these in their food, it is interesting to find out just what they need. Recently a copepod, not a pelagic but a rock-pool species, has been reared quite free of bacteria and has been fed on cultures of a number of different flagellates, themselves also bacteria-free. Not one of the flagellates tried so far has by itself been an adequate food for the copepods. When fed on one species only they will grow up, mate and lay eggs; and the eggs will develop to adult and the cycle be repeated several times, but eventually, sometimes after only one or two generations, sometimes after 15 to 20, the copepods will stop reproducing and die. It seems that they have a store of some essential substance in their body which is large enough to last through several generations but eventually gets used up. The substance or substances must come from the food. The interesting thing is that there are pairs of flagellates which, when given together, will support life indefinitely; one species must supply what the other lacks. It is therefore the fact that a copepod gets a mixed diet in the sea that keeps it healthy.

By using radioactive tracers it has been found that *Calanus* digest their food rapidly. The copepodite stage V builds up the bolster of fat which lies along the gut; in adult females food goes largely to the ovary and eggs, and indeed in an actively laying female up to three quarters of a given meal can reappear as eggs within a week. The number of eggs a copepod lays usually depends directly on the abundance of the phytoplankton (Fig. 29). Considering also the effort they must make to keep themselves up in the water against the force of gravity, and the extensive daily migrations they undertake from deep water to the surface and back, they must use up a great deal of energy and need a lot of food.

The usual way of measuring respiration is to put one or more copepods in a bottle of sea water whose oxygen content is known, leave them for a given time, and measure the oxygen again. Since bacteria are always present in sea water and multiply so actively when enclosed in a bottle that their own respiration affects the results, antibiotics are now usually added to the sea water, or else the period of experiment is made so short that bacteria have not time to multiply. One adult *Calanus* in about 10 ml of sea water will use up enough oxygen to be measured satisfactorily in one day, but, so as not to cramp the movements of the animal too much, it is usual to take a larger bottle and either a longer time or else correspondingly more *Calanus*. Because of the difficulties which would ensue if a diatom culture were producing oxygen at the same time as the copepods were using it up, it is customary to do the experiment in filtered water, i.e. under starvation conditions. The figure we get for oxygen used therefore covers only maintenance. In the experiments which have been made to compare feeding and starving copepods it was found that those fed used more oxygen than the unfed. In the sea in summer the young copepods will be growing and moulting and the adults will be producing eggs and sperm, and both will be using more than a minimum amount.

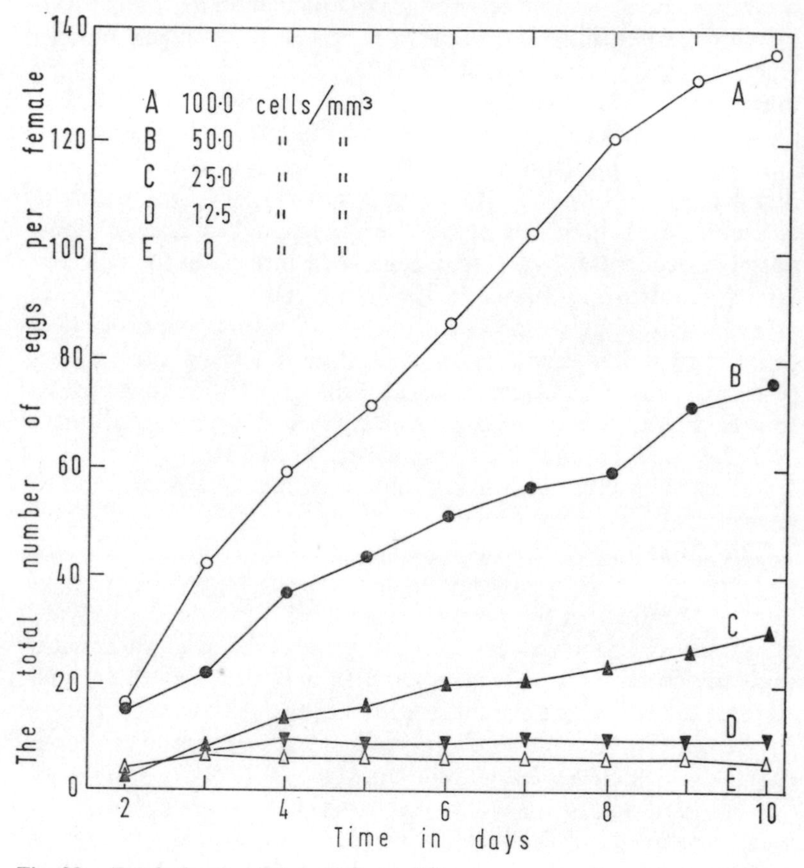

Fig. 29.—Egg-laying by *Calanus* fed on different concentrations of the diatom *Skeletonema*.

Respiration measurements have been made on some twenty species of copepod of differing size. As one would expect, the larger use more oxygen than the smaller, but perhaps a more informative measure is the amount of oxygen used per unit weight. In this case the small copepods are seen to be the most active metabolically, for they use 20 or 30 times more oxygen per unit weight than do the large. It is a general rule that metabolic activity decreases with increasing size, and it means that a small animal requires relatively more food than a large.

One fact which must be remembered is that all these respiration figures are obtained from experiments in the laboratory, where the conditions are very different from those in the sea. The copepods are often crowded and are enclosed within rigid walls, so that they are restricted and probably bump against a solid object much more often than in the sea, where

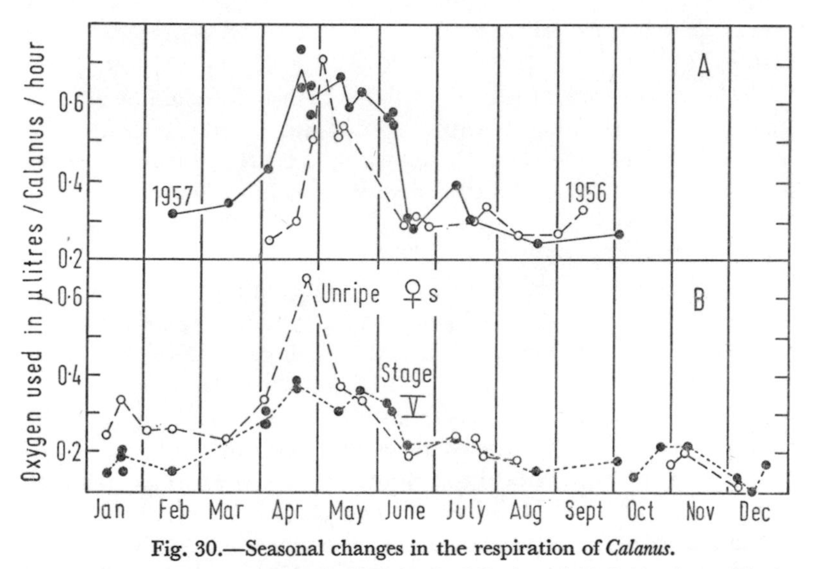

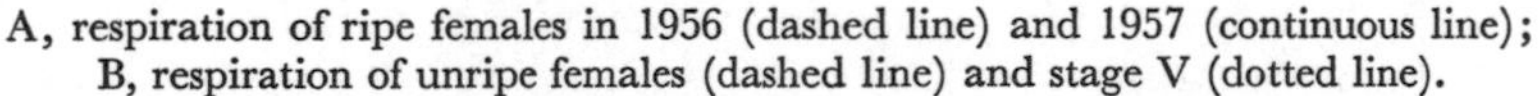

Fig. 30.—Seasonal changes in the respiration of *Calanus*.

A, respiration of ripe females in 1956 (dashed line) and 1957 (continuous line); B, respiration of unripe females (dashed line) and stage V (dotted line).

they have virtually infinite space. In the sea an observing diver has said that often the majority of the *Calanus* hang motionless in the water. This is not true of a catch of *Calanus* brought into the laboratory, and it means that our measurements of oxygen used and energy expended may be too high.

Measurements of *Calanus* respiration have been made over the course of a year and were found to vary with the season. In the spring months of April, May, and June, sometimes for a few weeks only, sometimes for 8 to 10, the respiration was much higher than during the rest of the year, at the peak about three times as high. This is the time of the first generation, the largest of the year, and also a time when the sea is usually rich in phytoplankton (Fig. 30).

In winter when respiration is low the population is all stage V and lives in deep water without diurnal migration. Its activity is therefore likely to be low and its requirement for food at a minimum. If we calculate, as protein, the proportion of its body weight which a *Calanus* will need to maintain itself it comes to about 7% for a female and, for a stage V, to about 3%. It is slightly more in summer, slightly less in winter, and the fact that *Calanus* is heavier in summer and lighter in winter increases the difference in actual food requirements. To express these quantities in terms of phytoplankton cells is difficult but it has been calculated that the amount a *Calanus* requires is contained in about 70 ml of fairly rich sea water. In lower latitudes, with winter diatom increases, the picture may be different.

This is to look at the problem from the point of view of the amount the copepod needs. We also want to know what a copepod actually does take in. Some diatom and dinoflagellate skeletons can be recognized in the faecal pellets and by counting them it is possible to tell how many have been eaten. The plants whose skeletons remain whole and countable are however very few indeed and this method is not really practicable. The simplest way is to put one or more copepods into a known concentration of phytoplankton for a given time and then to remove the copepods and count the phytoplankton left. Some precautions must be taken when using this method. The vessel must be large enough for the copepods to swim freely; the phytoplankton must be kept stirred, for if it settles to the bottom the copepods will get a richer supply there; the concentration of the phytoplankton must be arranged so that the copepods remove a good fraction of it, one quarter or more, otherwise the difference between the counts will not be big enough. Instead of counting cells, some constituent of the phytoplankton, such as carbohydrate or chlorophyll, can be measured.

Another method is to label the phytoplankton culture with a radioactive tracer, usually radioactive phosphorus, ^{32}P, and, as before, to put a copepod in a bottle of the culture for a given time. The number of cells eaten and digested can be measured by the radioactivity of the copepod body. We can assess the digestion by collecting the excreta of the copepod and measuring their radioactivity. This method is very delicate because it can measure minute quantities, but it does not measure liquid excretion and under the conditions of the experiment this can be quite large, so results by counting cells are usually higher than results with radioactive phosphorus.

Recently a chemical method was used. A number of *Calanus* were kept in a vessel through which natural sea water was flowing, and the organic matter was measured in the water before it entered and after it had left the vessel. In this way the amount taken out by the *Calanus* could be calculated.

All these methods have their advantages and drawbacks and they lead to rather divergent results. Even when more than one method is tried in the same experiment the results are often very different. The cell count method usually gives the highest figures, the radioactive tracer method the lowest. The first suggests that a *Calanus* can clear of cells up to 250 ml of sea water per day, the second that it rarely deals with more than 100 ml and very often with only 5–10. The chemical method showed that the needs of the *Calanus* could be covered by clearing 10–36 ml per day. A great deal must depend not only on the condition of the copepod but on the condition of the culture on which it is fed. Old cultures seem to be less easily digested.

The mathematical models set up to explain the spring diatom increase were mentioned in Chapter 3. In them, after allowances had been made for natural mortality and the carrying of diatoms below the photosynthetic zone by water movements, the reduction in diatom numbers was attributed to grazing by the zooplankton and the numbers of these were counted. It turned out that to account for all the diatoms lost between one visit and the next, each *Calanus* would have to sweep clear from 1–3 litres of water per day. This is a figure very much higher than anything found for *Calanus* in the laboratory, and various suggestions have been put forward to explain it, such as that *Calanus* does not behave normally in a laboratory experiment, or that it destroys a great deal more phytoplankton than it takes in, or that predation on the *Calanus* themselves has not been sufficiently allowed for. Only further research can solve the problem.

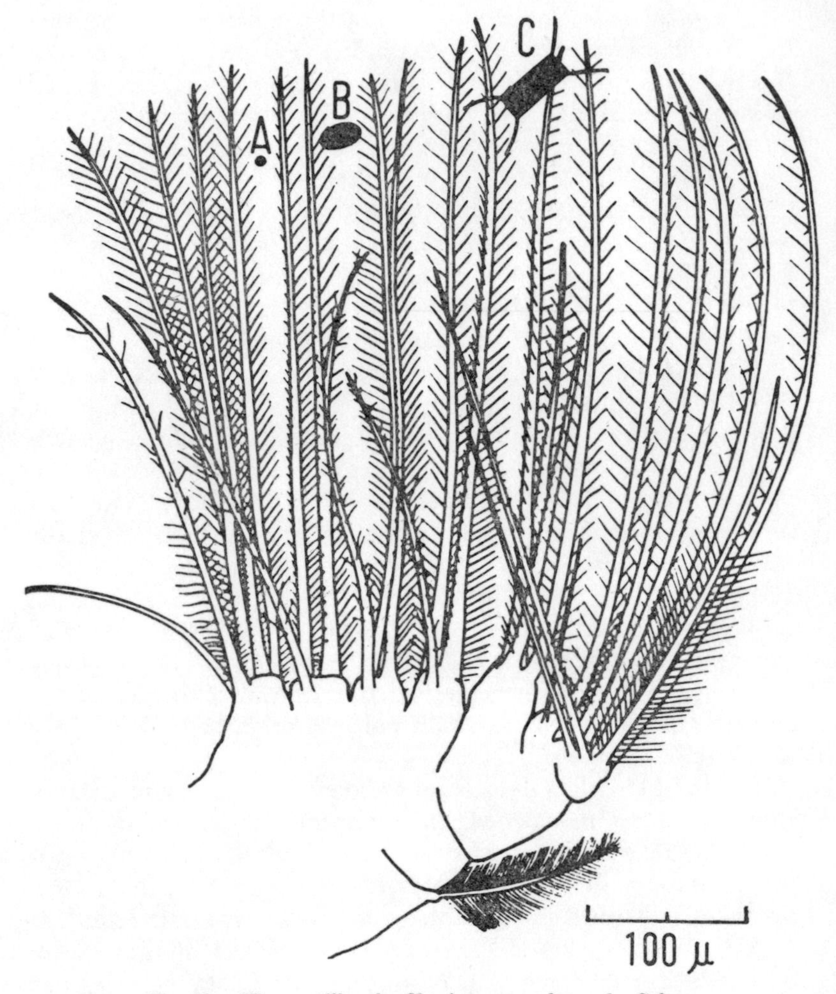

Fig. 31.—The maxilla, the filtering appendage of a *Calanus*.

A, B, and C represent three phytoplankton cells of different sizes to compare with the mesh size of the filter. A, a μ-flagellate; B, a coccolithophore; C, a diatom.

Feeding experiments have been done also with a number of the small copepods, either by the cell count or by the radioactive phosphorus method. As one would expect from their respiration they can filter off relatively more than *Calanus*. To use the word "filter" is perhaps misleading, for although they do have filtering mouth parts and use them as a rule they are also predatory at times or can select and eat large cells.

If we look closely at these filtering mouth parts we find that the food organisms are caught on a meshwork of fine bristles. The distance apart of these has been measured; it varies a good deal, but the smallest of the phytoplankton cells slip through them (Fig. 31) and on the whole are not available as food to copepods. Anything below one hundredth of a millimetre is not efficiently filtered off by *Calanus*, and surprisingly enough this is true also of young stages of *Calanus* and of many of the smaller copepods as well.

Of course there are zooplankton animals which are able to cope with the tiny cells. The Appendicularia (see p. 68) can do so, and in coastal waters the larvae of bivalves such as mussels and oysters depend on these very small plants.

The importance of all these figures is that they give us a measure of what effect the zooplankton, or at least the copepods of the zooplankton, will have on the phytoplankton in the sea. As we have seen (p. 47) this is not confined to predation, for their excretion of nitrates and phosphates has a fertilizing effect on the phytoplankton. During the summer when zooplankton is always fairly abundant the phytoplankton is practically all eaten and none sinks to the bottom to feed the population there. It may be very different in the spring; furthermore, the spring diatom increase may be one of two types. If the water layers are stabilized early, before there is much zooplankton, then the diatoms get a good start and there is an explosive outburst of growth. The number of diatoms is more than the few zooplankton can deal with and most of them sink out of the surface layers unused, to be eaten by the bottom-living animals or to form a layer on the bottom. In years when water is stabilized late, or if a stable layer is formed early and is broken up again by stormy weather (see p. 45, Fig. 15), then the zooplankton has time to begin reproducing. Under these conditions the zooplankton keeps pace with the phytoplankton, most of the latter is eaten, and the number of diatoms does not rise very high.

In regions of up-welling (p. 45), where nutrient salts are brought to the surface, the phytoplankton increases and this in turn starts off a zooplankton increase, but because zooplankton develops more slowly than phytoplankton it will often be carried away by the currents and will not appear in great numbers in the same place as the phytoplankton. Thus phyto- and zooplankton populations may sometimes appear to be separated in space, although the second has really been dependent on the first for its existence. The volume of catches of the larger plankton animals was measured on the long voyages round the world of the Danish research ship "Dana"

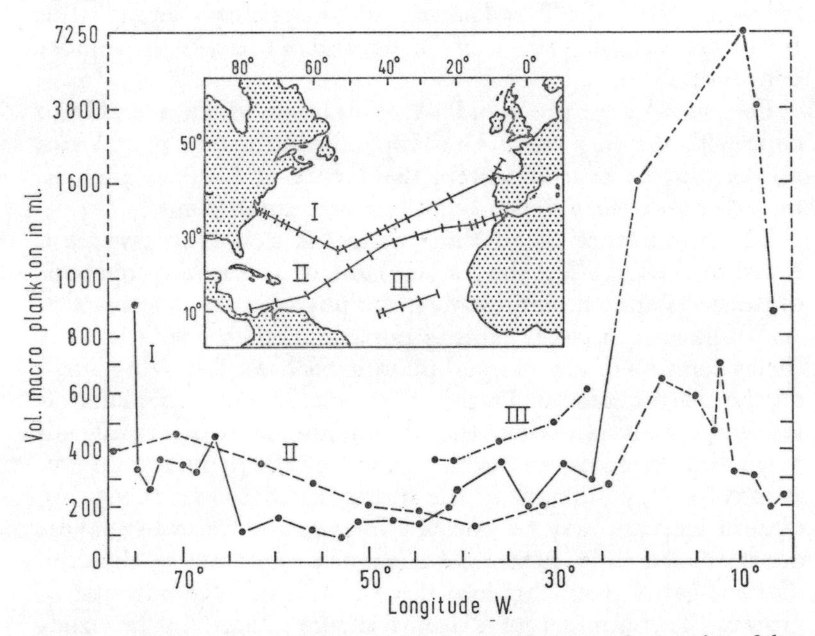

Fig. 32.—Volume of macroplankton taken by stramin net on three cruises of the research vessel "Dana" across the Atlantic. Inset, map showing the course of the cruises, with positions of sampling marked by cross lines. (*After Jesperson*)

(Fig. 32). This volume was usually higher at the surface than in deep water, and was poor in the middle of the oceans but rose both towards the coasts and in more northern latitudes.

At times of the year when phytoplankton is scarce it is difficult to know on what the zooplankton exists. In winter *Calanus* is present as copepodite stage V, which, as already mentioned, uses less oxygen than the adult, stays in deep water and does not migrate vertically, so saving energy. At such times too copepods are probably predatory, even cannibalistic. The life history of one large arctic species of *Calanus*, *C. hyperboreus*, has been investigated at Woods Hole, Cape Cod. This copepod lays up a large store of fat (up to 70% of its body weight) by feeding in spring and does not feed at all in late summer and winter, during which time its respiration decreases and it lives on its fat reserves. Its specific gravity decreases and it floats in the water, hardly moving. This curious life history, with the animal living and feeding actively for less than a quarter of its lifespan, is perhaps an exaggerated version of what occurs in other copepods.

A possible source of food is detritus, the broken-down fragments of dead plants and animals or of animal excreta which remain for a while, slowly sinking in the water. Detritus is often more abundant than living plankton, but too little is known about its edibility or its distribution to assess its importance as food for zooplankton. It contains a certain amount of organic matter and indeed this was included in the estimation of food available in the experiments on feeding *Calanus* in running water (p. 89), but as yet nobody has succeeded in feeding copepods on detritus alone. The quantity of dissolved organic matter is even more than that in particulate form, but seems to be unavailable to the copepods. It has been discovered however that when sea water is shaken up so much as to form air bubbles, as it must often be in stormy weather, the organic matter comes out of solution and forms little flakes on the bubbles. The brine-shrimp *Artemia* can feed on these little flakes of material and it seems a possibility for copepods too.

So far we have been dealing with zooplankton living in the top 100 or 200 metres. There are also animals living much deeper, in water down to 6000 metres, and their nutrition is an interesting problem. The plants and animals dying in the surface layers disintegrate fairly quickly and few of them reach the great deeps in a form fit to be eaten. There are a few abyssal detritus feeders, but not many. A surprising number of zooplankton animals perform enormous vertical migrations of several thousand metres to feed in the surface layers, and they are the main source of food for the non-migrating deep-living animals. Some of these, indeed, do a reversed migration, coming up by day to meet the surface feeders when they return from the upper layers. The faecal pellets of the surface feeders must often be produced in deep water during the day and will be a source of food for the detritus eaters. Much work remains to be done on the relations between the predators and herbivores of these deep waters.

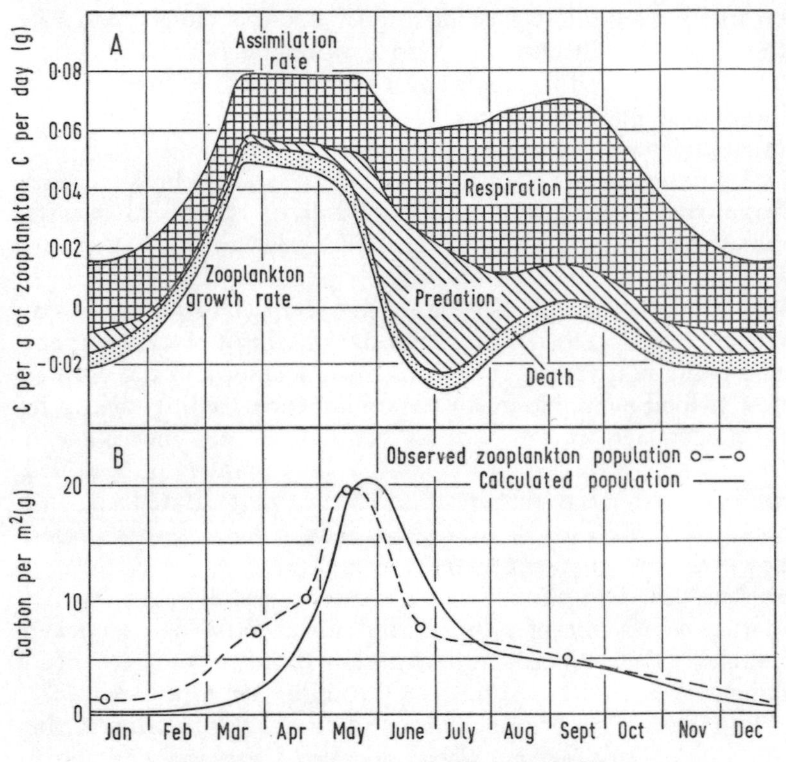

Fig. 33.—Model of a zooplankton population.

A, summation of zooplankton growth processes; B, observed seasonal cycle and the theoretical population curve calculated on the basis of seasonal variation in environmental factors. (*From Riley*)

A mathematical model has been constructed for the zooplankton cycle in the same way as for the phytoplankton. The factors which have to be taken into account here are the growth of the phytoplankton, the rate of growth of zooplankton, its respiration, its natural mortality, and its removal by water movements and by predation. The model (Fig. 33) was made for Georges Bank in the Gulf of Maine, where the most important predator was the arrow-worm *Sagitta*. The respiration and rate of feeding of the herbivores were based on experiments on copepods. As with phytoplankton the model fits the actual cycle moderately well, although the rate of growth in the model is too low in spring and too high in early summer. Similar equations have been put forward for the North Sea where in 1954 it was possible to follow the same patch of *Calanus* from visit to visit over a period of about 10 weeks.

We have seen that the phytoplankton can use only a small fraction of the energy that reaches the sea as light. We must now look at the efficiency of the zooplankton in converting the phytoplankton into fish-food.

Intake of food = Respiration + Excretion + Growth

The first item in this equation is the total amount of food that an animal takes in, the respiration represents the energy used up in daily needs, such as feeding and swimming, the excretion is that part of the food taken in which cannot be digested, and the last item covers not only the growth of the individual but also any eggs and sperm produced. Each item can be expressed as units of energy, and the efficiency of conversion is the proportion of food taken in which is used for new growth. This is the gross efficiency; the net efficiency is the fraction of the *digested* food which is used for new growth.

One of the fullest investigations that has been made was not on a marine animal but on a freshwater crustacean, the water-flea *Daphnia* (related to the *Podon* and *Evadne* of Fig. 21(G,H)). Populations of three different ages were kept in the laboratory and fed on four different concentrations of food culture. The younger groups were pre-adult but became adult during the course of the experiment. The length, weight, feeding rate and respiration were all measured either during the experiment or on similar populations. Although the amount of food consumed rose as the concentration of food increased, yet the proportion which was digested decreased, so that more food was wasted and the gross efficiency was less. Nevertheless the proportion of the digested food which was used for growth and reproduction (net efficiency) increased slightly. The numbers of young hatched in the high food concentrations were always greater than those from the poor ones. The gross efficiency ranged from

4–13% and the net efficiency from 55–59%. Up till the time when the *Daphnia* became adult most of the digested food went to growth, and after that, to eggs. This is reminiscent of what happens in *Calanus* (p. 85).

From a general consideration of the quantities of phyto-plankton and zooplankton in the sea it has been concluded that copepods are very efficient converters and that their net efficiency is about 70%.

That this high figure is sometimes possible is shown by a detailed study of a euphausiid shrimp off the coast of California. It was done in a series of one-day experiments by feeding the *Euphausia* with a culture labelled with radioactive carbon, measuring at the end the culture, the shrimp, the oxygen used, and the excreta. There were great differences between the eight animals tested: net efficiencies varied from 11–74%, and the average was 32%.

It is obvious that on the whole the conversion from phyto- to zooplankton is an efficient process when compared with primary production.

Chapter 6

Some Food Chains

Third Fisherman—Master, I marvel how the fishes live in the sea.
First Fisherman—Why, as men do a-land, the great ones eat up the little ones.
W. SHAKESPEARE, "Pericles", Act III, Scene 1

Much has been said in the preceding chapters about the food links between the phyto- and the zooplankton and the food chains which link the nutrient salts in the water with the fish and whales which are caught for man's use. We shall now describe a few examples of these in more detail.

Food-web is perhaps a better description than food-chain for the complicated relationships between plants and animals in the diet of the herring as shown by Hardy (Fig. 34). The herring is our most valuable plankton-feeding fish. It lays its eggs on a suitable surface on the sea bottom and they hatch in a week or two (depending on the temperature). After the young have consumed the yolk in the yolk sac, which lasts them only a few days, they have to fend for themselves and this is a critical period in their life. They can take in both plants and animals but these have to be very small because of the size of the larval gullet. If too large an animal is eaten it sticks in the gullet and the young larva dies. Moreover, one observer has noticed that they soon get discouraged and if they fail to get anything after two or three attempts they give up and die. Each organism is taken by a definite act of capture and not by chance encounter (Fig. 35). The larvae therefore cannot feed in the dark, and in the sea maximum feeding is in the early morning and late evening.

First the larvae take diatoms (although it is not certain that

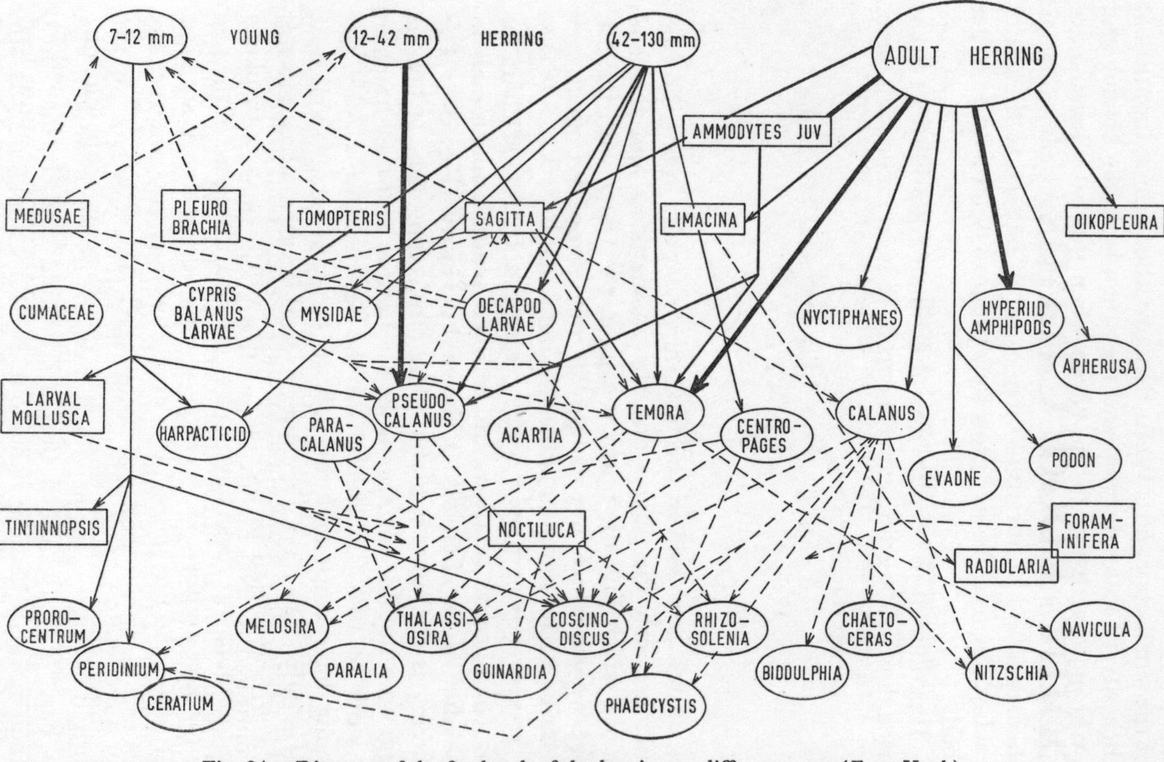

Fig. 34.—Diagram of the food web of the herring at different ages. (*From Hardy*)

they can digest these) and the nauplii of the small copepods. As they grow larger they can take the copepodites and adults of the small copepods. When they are about three months old their silvery scales develop and they then look like herring and swim more actively. In the Clyde when the baby herring are nearly three months old the *Calanus* of the first generation lay their eggs and hundreds of these eggs can be seen in the gut of each young fish. As they grow they progress to larger and larger food.

It is difficult to keep herring alive in an aquarium but it has been done and feeding experiments have been made with them.

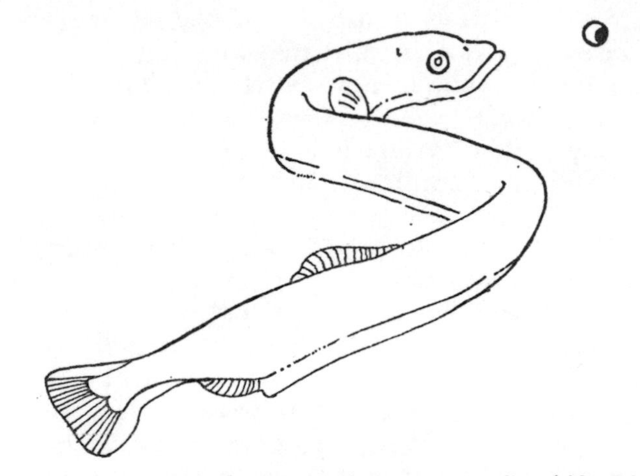

Fig. 35.—Larval herring just about to capture a nauplius. (*After Soleim*)

When they were given a good supply of copepods (about 9 per litre) they broke shoal formation and began to feed. They continued to feed for about two hours and then the shoal re-formed. The rate of digestion depended on size and temperature. For a fish of six inches complete digestion took 24 hours. Even at this size they still cannot feed in darkness and do best in a moderately good light. Near the surface even moonlight is bright enough—"the herring it loves the merry moonlight" has a basis of truth. When they are feeding actively they break their normal shoaling behaviour and dart about rapidly after the food.

A careful study of the food of the herring has been made in

the North Sea (Fig. 34). Nearly half was crustacean; in this fraction copepods and amphipods were equally abundant and there were also a few euphausiids. Most of the other fraction was made up of fish (sand eels) but there were a few *Spiratella*, arrow-worms, and appendicularians (Fig. 23(L)). The herring on the west coast of the British Isles eat a much higher proportion of crustaceans, mainly *Calanus* and euphausiids. Most of these food organisms, as we have already seen, feed on the phytoplankton, but the arrow-worms and some euphausiids are themselves carnivorous and will feed on copepods, and the first even on small herring larvae.

The plaice is an example of a bottom-living fish. Its eggs however are laid freely in the sea and the larvae spend some weeks in the plankton. At first they look just like round fish, but soon one eye begins to move over to the other side, the body flattens from side to side and the little fish settles on one side on a sandy bottom. While it is in the plankton it feeds on plankton organisms suitable for its size, usually copepod nauplii, small copepods, larval molluscs, and worms. In the southern North Sea, however, it depends almost entirely on the appendicularian *Oikopleura*, which itself feeds on the very minute flagellates of the plankton (p. 68). Once on the bottom the plaice changes over to feeding on bottom-living animals such as amphipods, worms and bivalves, which in turn have a variety of ways of feeding. Many bivalves live buried in the sand, connected with the surface by two tubes, the siphons, one of which takes in water and the other discharges it. In the bivalve *Scrobicularia* one of the siphons is extended into a long thin tube which moves actively about over the surface (Fig. 36) like a vacuum cleaner, sucking up the detritus. In this detritus there are many little bits of organic matter, some living, such as bottom diatoms, small nematode worms or bacteria, some the dead remains of plants and animals which have sunk from the surface layers, for in shallow coastal water these will reach the bottom before dissolving. *Scrobicularia* is a deposit feeder, but there are many burrowers both molluscan and crustacean which are suspension feeders and live on the small planktonic plants and animals.

Apart from being commercially valuable the plaice is a hardy and easily handled fish and so a good deal of work has been done on its feeding and growth. A number of plaice were kept in boxes floating in the sea and were fed with known quantities of mussel flesh. Some were given just enough food to keep them at a steady weight—a maintenance ration; some were given more than this—an intermediate ration, and some were given as much as they would eat—a maximum ration. To keep a plaice of 40 g (about an oz. and a half) at a steady weight without growth needed between 4 and 10 g of food a week, or between $1\frac{1}{2}$ and $3\frac{1}{2}$% of its own weight per day. If the water is

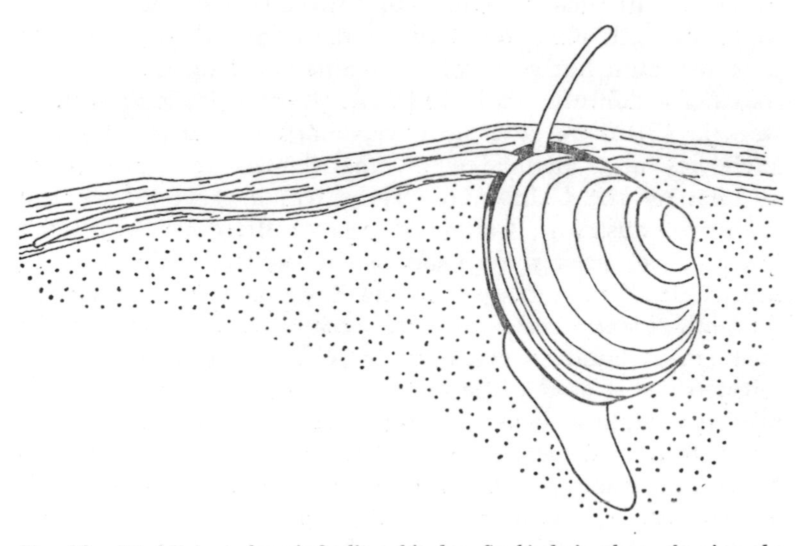

Fig. 36.—Mud-living, deposit-feeding, bivalve *Scrobicularia plana*, showing the mode of feeding. (*From Yonge*)

warm it needs more, if cold, less. Bigger fish need proportionately less to maintain themselves. With a larger food supply they will grow, but with twice the maintenance ration they grow almost as fast as with a maximum ration. With fish fed on the maximum ration only a quarter or a fifth of the food was used for growth.

Whalebone whales depend on plankton for food, although some species take small fish as well. They filter off the plankton through the fringes of whalebone which hang inside their mouths and, since they need enormous quantities to maintain themselves, they are found most commonly where zooplankton is richest, in the Arctic and Antarctic. The commercially valuable whales of the Arctic have been destroyed by over-fishing, and the populations of the Antarctic are rapidly going the same way. The food of the northern whalebone whales is largely *Calanus* and *Spiratella*, that of those of the southern ocean mostly *Euphausia superba*. As we have seen these foods themselves exist on phytoplankton.

There are not many herbivorous fish but among them are some of the herring family, sardines and anchovies. One curious feature about their feeding is that the larval stages eat more copepods and other zooplankton than the adults; both eat phytoplankton but some of the adults eat little else. Off the coasts of California, Chile and Peru where there is up-welling (see p. 45) there are, or were, enormous shoals of fish—anchovies and menhaden off the South American coast, sardines off the Californian. There was a great fishery and canning industry in California (immortalized in Steinbeck's *Cannery Row*) until some years ago when the sardines disappeared. In Peru nobody seems to have thought of using the fish until recently and innumerable sea birds, the guayanes, fed on them. A hundred years ago the chief source of nitrate, or salt-petre as it used to be called, was in Chile where, in this rainless area, huge beds of "guano" have been formed over the centuries from the excretions of these sea birds. Until the mid-1950s there was only a small fish-canning industry in Peru, and a small fishmeal industry depending on its residues. With an increasing demand for fishmeal the latter industry was expanded, processing machinery was brought south from the derelict Californian factories, and a small-meshed purse seine was introduced. This caught the "anchoveta", a small fish of the sardine family, 10–17 cm long, which occurs in countless millions off the Peruvian coast and is the chief food of the guayanes.

These fish are not canned; they are cooked, dried, and ground to a meal which is used as a high-grade supplement to foods for pigs and poultry. Besides meal, oil is recovered and used in the manufacture of paint and linoleum. A third product is a liquid residue containing nutrients and this is also sold as a stock-food supplement or as liquid fertilizer.

The industry rocketed. In 1948, 1,000 metric tons of fish

were landed; in 1956, 138,500; in 1962 and 1963, more than 6½ million tons. With calm weather and the fish supply close to the shore, even unskilled fishermen can fill their holds once or twice a day. The resulting meal can therefore undersell European fishmeals.

Every few years however the nutrient-rich Peruvian current water is overlaid by a warm sterile water mass flowing from north to south, a current which has been given the name "El Nino". When this happens the fish are no longer found near the surface, the guayanes are no longer able to catch their usual food and many starve to death. Such bad years have been recorded ten times in the last half-century; the last was in 1957 when the guayanes are supposed to have been reduced from 35 to 10 million. Something of the same sort happened in 1963 (although it was not caused by El Nino) and although the total catch remained high there had been an increase in fishing effort. After this comparatively poor year the catch consisted largely of small (8–11 cm) and immature fish. It seems probable that the stock is now being overfished.

The production of guano has decreased steadily since the anchoveta fishery began, and although El Nino has always caused fluctuations in the number of guayanes, the guano producers now blame the decrease entirely on the anchoveta fishermen. It has been estimated that the birds have to eat 11–16 tons of fish to produce one ton of guano, and since a great deal of their excretions must be lost at sea (thus increasing surface fertility) a conservative estimate would be 22 tons of fish for one ton of guano, whereas 5–7 tons of fish produce one ton of fishmeal besides the oil and fish solubles.

This is one of the biggest fisheries in the world, taking about 18% of the total world catch, and conservation measures are obviously called for.

There is a commercial herring fishery in the Barents Sea and recent work has shown an unexpected and intricate relationship among the fish and zooplankton. In winter, when *Calanus* is in deep water, the herring are in deep water too and they follow the *Calanus* up to the surface in spring when these spawn, feeding on them actively. In some years however the ctenophore *Bolinopsis* is so abundant that it destroys the *Calanus* and spoils the herring fishery. As we have seen, ctenophores eat herring larvae and those of other fishes was well as copepods. Russian scientists have found that one ctenophore $2\frac{1}{2}$–3 cm long can eat 10–11 copepods in two hours, digesting half of them in one day and throwing the rest out dead. With *Calanus* at 300/cubic metre, and two *Bolinopsis* the *Calanus* would all be eaten in a month. However there is a larger ctenophore *Beroe* which does not eat crustacean food but lives on other ctenophores, often *Bolinopsis*; *Beroe* in turn is eaten by the cod, which sometimes gorges on it. Haddock eat *Beroe* too although not so intensively as the cod. In years when cod are abundant in the Murmansk coast fishery they reduce the population of *Beroe*, thus increasing the population of *Bolinopsis*, and so the stocks of *Calanus*, and then of herring, suffer. A good herring fishing may therefore depend on a bad cod fishery, and this through a whole series of inconspicuous links.

Some of the herbivorous zooplankton, if themselves useless as food, may be in direct competition for food with the useful species and so have an injurious effect on fish populations. Salps may be an example of this but they are not often common in northern waters.

Breakdown and regeneration are part of the cycle of life and this is where bacteria are important. When a plankton animal dies its substance begins to decompose at once and within a day or two the body has lost almost all its nitrogen and phosphorus. This does not seem to be by bacterial action; probably the enzymes present in the body can account for it. We have already seen (p. 44) how important it is for the productivity of the phytoplankton in summer that dissolved nutrients should be returned to the water quickly.

The remains of the animal bodies and of the faecal pellets, which are not dissolved quickly, sink to the bottom (if they are not eaten on the way) and are there decomposed by bacteria and gradually returned in solution to the sea water.

Bacteria such as those shown in Plate VIII are found throughout the oceans, but they are much more common on the bottom than elsewhere, particularly on the surface of the bottom sand and mud deposits. Here they break down dead animal matter and here they themselves are a rich food for the bottom-living detritus feeders. In the water itself they are much scarcer and become rare as one goes towards the open ocean.

Many bacteria are washed into the sea from the land and fresh waters, but these do not survive for long. Even the dangerous coliform bacteria and typhoid bacilli are killed within a day or two. If, however, they are eaten by shellfish such as mussels or oysters they can survive in the body fluids of their hosts, and this is why there are such stringent regulations about the cleansing of those shellfish taken for human food. Marine bacteria are however adapted to salt water and can even withstand the pressure at great depths. Although in a sample of water taken from the open sea very few bacteria can be counted, yet if this same sample is put in a bottle the numbers rise immediately. The smaller the bottle the greater the number that develops. They are living upon the very dilute solution of organic matter in the sea water but they also seem to need a surface to settle on, and the greater the surface the higher the number. Putting glass beads in the bottle increases the surface available and raises the numbers.

The reason why these bacteria do not multiply while they are in the sea is not fully understood. Possibly it has something to do with the extreme dilution of their nutrient medium, the sea, for if nutrients are added to the sample bottles the difference (in numbers developing) between large and small bottles disappears. It has been suggested that strong light at the surface is injurious to them, but this effect could not go very deep. It may be that in open water the bacteria are all attached, either to plankton organisms or to particles of detritus.

It is for these reasons that most bacterial decomposition takes place on the sea bottom and there the dead material is eventually broken down into nitrate, phosphate and carbon dioxide which are again available for phytoplankton growth. If the sea is shallow these nutrients will be returned to the surface waters with the temperature overturn in winter (see p. 15). If it is very deep they may be lost to the productive cycle unless they are carried along in some deep bottom-flowing current, to well up to the surface in places far from their origin.

Bacteria are of very many types and have different functions. There are some which are able to decompose tough substances like chitin, cellulose or lignin. There are others which oxidize ammonium compounds to nitrite and nitrite to nitrate. This is the usual sequence, but in the sea there are also bacteria which do the opposite and reduce nitrate to nitrite or even to nitrogen gas. They can use all these substances for building up their own bodies and in addition they produce enzymes and even vitamins which are of vital importance to the growth of the phytoplankton. We have seen that it is much more difficult to grow cultures of phytoplankton without bacteria than with them, but the substances they supply have not always been identified.

Chapter 7

Production of Fish and Shellfish

> —"and even so, although one acre of ground overflowed with water will naturalie and of itselfe keep but three or four hundredth carpes, or other fishes, yet so much feeding as you may adde there to, that it may keepe three thousand or four thousand in as good a plight as three hundredth or four hundredth without such feeding".
>
> J. TAVERNER, 1600, "Certain experiments concerning Fish and Fruites".

The energy on which depends the cycle of life in the sea comes from outside the sea—it is the energy of the sunlight falling on the surface. The whole system may be seen as a series of nutritional or trophic levels, each depending on the level below and in turn providing sustenance for the level above. Fig. 37 is a diagrammatic representation. The seaweeds and phytoplankton, the producers, use sunlight to build up, from the simple salts dissolved in the sea water, more complicated proteins, fats, and carbohydrates in their own bodies. These producers are eaten by the primary consumers, which in the open sea are zooplankton and on the shore are the browsers and the detritus- and filter-feeders. These in turn are preyed on by the secondary consumers, predators such as *Sagitta*, euphausiids or jellyfish in the sea, carnivorous worms or gastropods on the shore. Finally come the fish or tertiary consumers, some of which are bottom feeders and some planktonic. Of course, the food chain may be lengthened by the intervention of another set of predators, fish feeding on fish for example, or shortened by leaving out one trophic level as when fish feed directly on zooplankton. The bacteria can be considered as decomposers and transformers, dealing with the excreta and dead bodies of all levels.

The amount of energy which one trophic level passes on to the next is bound to decrease with each level. For one thing,

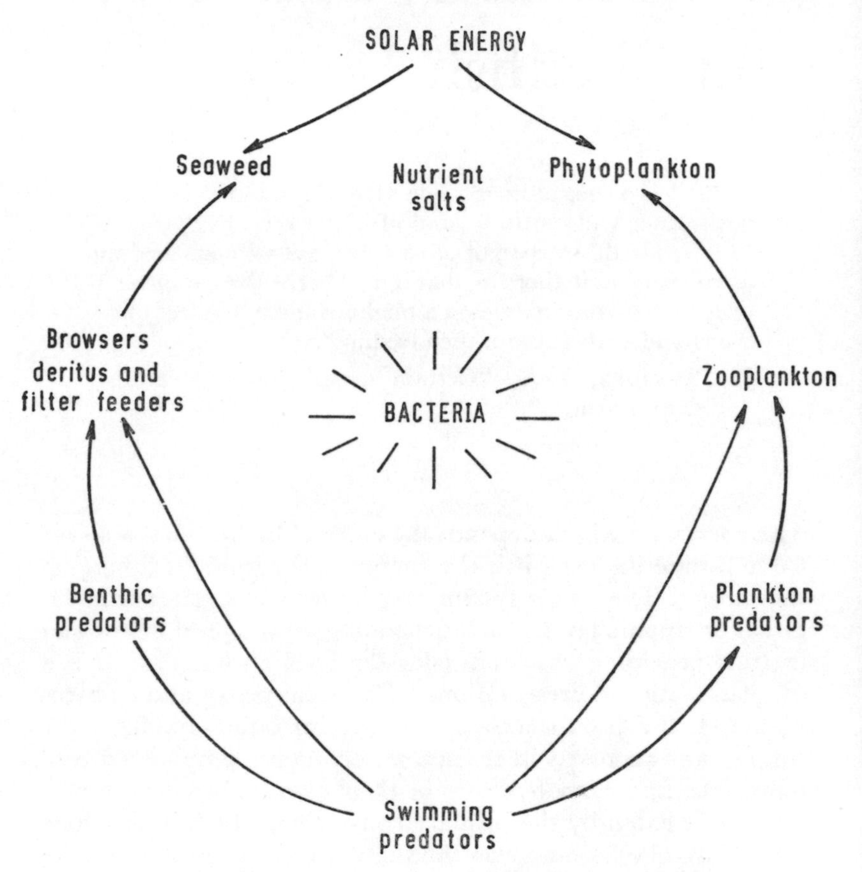

Fig. 37.—Diagram to show the cycle of production in the sea.

there are those plants or animals which are not eaten by the predators of the level above but survive to carry on the stock, and secondly, there are those which die a natural death and whose organic substance is returned by autolysis or by bacteria directly to the dissolved nutrients in the water. Furthermore, an organism cannot use all its food for building up new body tissue—it uses some for its own activities such as swimming and feeding, and this fraction can be measured by its respiration. We have seen in Chapter 5 that:—

$$\text{Assimilation} = \text{Respiration} + \text{Growth}.$$

It is obvious that if respiration increases the efficiency $\left(\frac{\text{growth}}{\text{assimilation}}\right)$ will decrease. As one goes from one consumer level to the next the respiration does tend to rise, and the efficiency drops. The efficiency of the zooplankton was estimated at 70% (p. 96); that of the fishes was put by the same worker at 5–10%. This is understandable when one thinks of the relative activities. Phytoplankton is almost motionless, zooplankton has but feeble powers of movement, and fish swim actively to catch their prey.

The amount of energy passed on from one trophic level to the next is what we call the productivity, and although the most interesting from a practical point of view is the productivity of the top level, the fish crop, yet a study of the energy relationships at all levels will help us to understand and perhaps eventually predict this final one. Such work has been done more often in fresh water than in salt, mainly because a pond or lake is a more manageable body of water than the sea. In one very careful study of a not very productive American lake the quantities produced (expressed as g.cal/cm^2/year) were: for primary producers 111, for primary consumers 15 and for secondary consumers 3. For the sea the proportions have been estimated at 100, 33 and 0·4, but the figures must be taken as approximations only, for all these quantities are difficult to evaluate and the estimates depend on few and scattered observations.

There is one obvious way of increasing the efficiency of our food-getting, and that is to cut out one trophic level altogether. The experiment on growing *Chlorella* mentioned earlier (p. 50) is one way, but evidently not a very enticing one.

It was proposed during the war that zooplankton should be caught and used as food, particularly the copepod *Calanus* which is known to occur in enormous swarms. But these swarms are difficult to locate and the problems of filtering off the huge quantities of water involved were formidable, so that it is now considered easier to let the herring and the whale do the concentrating for us. However, as the Antarctic whaling industry is failing through overfishing it may become necessary to harvest instead the *Euphausia superba* on which these whales depend.

Another large group of primary consumers is found in the bivalve molluscs and here there has already been a great deal of work on exploiting and improving natural populations. Oysters and (very much behind them) mussels, are the most valuable species, but many others are eaten (especially in America) and would repay attention. Usually cultivation involves no more than collecting the young and transferring them to suitable places for growth and fattening, but even this means a great deal of work.

Oysters breed in summer and their larvae have a planktonic life whose length varies with the temperature of the sea water but may be as much as three weeks. During this time they depend for food on the smallest of the planktonic algae, the μ-flagellates. Two things are important for a good settlement of the young oysters or "spat"—the amount of food available and a clean surface to settle on. The second is the more easily provided, and in oyster fisheries a supply of dead oyster or mussel shells is put down on the ground each year. Other spat collectors may be suspended in midwater; these may be ropes, bundles of birch twigs, wire-netting bags full of shell, or tiles and cardboard egg-boxes covered with a cement which can be removed later along with the settled oysters.

This kind of cultivation has been going on for centuries. The Romans imported oysters from elsewhere and attached them to ropes hung from a wooden framework (Fig. 38). These "ostriaria" were situated in places where the oysters would grow quickly. Oysters and mussels slung from buoys in the Bay of Vigo (Spain) are today a flourishing industry.

The French type of cultivation is seen in Plate IX; the lower picture shows cement-covered tiles, stacked in crates, which are being examined for spat, and the upper picture shows an enclosed oyster park where the spat are laid on the ground to grow and fatten after being stripped from the tiles.

The food supply is not so easy to control, for the number and type of minute algae vary a great deal from year to year, but some attempts have been made, by adding nutrients to the water, to increase their number.

One of the strangest places where oysters breed is at the head of some Norwegian fjords. Normally the water in such latitudes would be too cold for breeding to occur freely. However, at the head of these fjords there are small areas of water, the polls, which are connected with the main fjord only by narrow channels, and often communicate with it only at high tide.

Fig. 38.—Designs from Roman vases showing Ostriaria: above, Piombino vase; below, Borgiano vase (form of this shown on right, actual height 11 cm). The ostriaria (which are labelled) come between drawings of other buildings, a pier, and a lighthouse (pharos). (*From Yonge*)

Under such conditions a brackish-water layer forms on the surface of the poll and this acts like the glass of a greenhouse, letting the heat of the sun through by day but insulating the deeper salter layers from the cold night air. The temperature therefore rises quite high in summer (up to 30 °C) and if minute algae grow abundantly the saltwater layers are a suitable place for oysters to breed and larvae to grow. The adult oysters are suspended from buoys in wire cages and various types of collectors are put out for the spat to settle on. Two or three of these polls were studied for several years on end and it was found that in sunny years all went well and plenty of green and blue-green algae grew to feed the oysters. In wet and overcast summers, however, breeding was a failure; dinoflagellates and bacteria grew instead of algae and even the adult oysters might die of starvation. In such years the addition of some nitrate and phosphate made a big difference and encouraged growth of the right algae; in sunny summers such an addition made little difference.

At sea the fisherman is a hunter, primitive in aim but modern in equipment and skill. The fishermen of this country take annually from the sea around us roughly a million tons of fish, worth many millions of pounds. Together, the other countries of Europe take several times this amount. This enormous supply of food is obtained without the attention to grazing and breeding which for centuries has been the mark of the farmer.

It used to be considered that man's fishing efforts were puny compared with the vast resources of nature. An area like the North Sea, for example, seems, at least to the unhappy landsman crossing it, almost limitless in its extent compared with the scattered tiny trawlers which go to and fro on its surface dragging their trawls or nets. Many fishermen indeed believe that their efforts have a negligible effect on the fish population. Two world wars have however shown conclusively that when the North Sea is given an enforced rest from fishing not only are there more fish afterwards, but the average catch is heavier.

In the sea each adult female fish, say of plaice, produces hundreds or thousands of eggs, most of which hatch into baby fish. Most of these die off, leaving still a considerable number of small, rapidly growing fish. They are of little use for human consumption until they are perhaps 9 or 10 inches long, and there are regulations to enforce the use of a net with meshes large enough to avoid their capture and also to ensure that any small fish captured are not put on the market but returned to the sea. From 9 or 10 inches up we get good marketable fish. Really large fish are not economic, for they are growing very slowly and have to take relatively more food merely to support existence.

The present position is that overfishing has greatly reduced the stocks in certain areas, for example the North Sea. It has been recommended for these areas that the fishing effort should be considerably decreased—in the case of the North Sea to one half—and fishery scientists calculate that if this were done the yield of the fisheries would rise by 20%. In the Pacific halibut fishery a restriction of fishing did lead to an increase in the catch.

The sea, however, except for a variable and now much disputed strip round the coast, is open to the fishermen of all nations and to get international agreement on conservation measures is a Herculean task. Progress is, however, being made.

A number of years ago an interesting experiment was done in the North Sea. As a whole it has been overfished but in some areas, for instance close to the coasts of Denmark, Holland and Germany, there are too many small fish which have not enough food to grow quickly. Boat-loads of these small flatfish were transported in well-vessels to the Dogger Bank, where they were marked with a tag so that they could be recognized again, and liberated. Because of the plentiful food there they grew at a remarkable rate and were soon of marketable size. Unfortunately this promising scheme has gone no further.

Fish are not the only predators on the sea bottom; they have many competitors for their food, among them worms, starfish, brittle stars, and carnivorous snails. Also, fish which are not useful to us as food compete with those that are. It seems too that invertebrates are on the whole more voracious than vertebrates. Most fish eat about 5–6% of their own body weight daily during the summer and less in winter. Young invertebrates take in up to 25% of their own body weight daily, although the adults are more restrained and take about 15% when active. When we consider how abundant some of these predators are (15 starfish or 400–500 brittle stars per square metre of bottom in some places, see Plate X) it is astonishing that any of the food organisms can grow up at all. There is, however, a period of relief for them while the invertebrate predators are breeding. Many marine organisms eat little or nothing for some time before spawning. This is true of starfishes and indeed in some brittle stars the ovary takes up so much room in the small disc-shaped body that there is no room for the gut, which disappears at spawning time and has to be reconstituted after the eggs are laid. Curiously enough the breeding period of the main predators comes at the time when their food has just settled and is growing rapidly. This benefits both prey and predator, for in a month a mussel, for example, will increase in weight 500 to 2000 times, which means that at the end of the month a starfish can for one meal eat one mussel instead of the thousand or so it would have needed at the beginning of the month, and can leave alive the other 999.

From investigations in the Kattegat it has been calculated that about a million tons of fish food have to be shared out among 5000 tons of flounders and 75000 tons of invertebrate predators. If we take into account the different rates at which they feed it seems that the fishes useful to us are eating only 1–2% of the food. Surely if more attention were paid to the destruction of the useless we could ensure a larger share for the useful. We should first have to make sure that what we consider useless really plays no part in the production of our fish.

Adding nutrient salts is the basis of any culture of microorganisms, whether bacteria, diatoms or flagellates, and it is constantly being done in laboratories. It has been tried successfully in seawater tanks of various sizes in the open air, and finally it has been tried in natural waters, both fresh and salt.

During the second world war an attempt at "fish farming" was carried out in a small loch on the west of Scotland—Loch Craiglin, about 18 acres in extent, opening by a narrow channel into one of the arms of Loch Sween in Argyll (Plate XI). It was shallow, being less than 2 m over most of its extent, with a deeper area of up to 5 m opposite the entrance of the channel. Sea water could enter the small loch freely at high tide, and a sluice gate was erected in the channel to control the inflow and outflow if necessary. Imperial Chemical Industries agreed to finance the experiment and supply fertilizers. The intention was to shut off the loch, add nutrient salts in the form of sodium nitrate and superphosphate, introduce oysters and young fish, and then study the development of the plankton, the bottom fauna and the introduced fish and oysters. The oysters should thrive directly on the increased phytoplankton and the fish on zooplankton and bottom fauna. Young plaice proved difficult to obtain in quantity and young flounders were used as well and in larger numbers.

Fertilizers were distributed in the loch at irregular intervals for the next year or two, with very varying results so far as the plankton was concerned. Sometimes an increase in numbers of diatoms or dinoflagellates would follow, sometimes nothing would happen. This depended largely on the time of year. During the spring and summer there was a rich growth of eel grass and green seaweeds round the shores of the loch and it became evident that these had first call on the nutrients. Fertilizers evidently increase the growth not only of desirable plants but also of undesirable, or "weeds". When the nutrients were added after the seaweeds and eel grass had died down in autumn, then the response by the phytoplankton was marked and in the second winter dinoflagellates rose to such extraordinarily high numbers that they coloured the water a reddish brown, and they remained abundant all through the spring.

Zooplankton was numerous during the first summer but not afterwards. However, the bottom fauna increased greatly, and continued to do so as long as fertilizations went on. It consisted largely of midge larvae (which are a good food for flounders), and small snails and cockles. The fish grew very well too and put on as much weight in one year as they would normally do in two or three. However, the loch became an odd place because of the repeated fertilizations. When the sluice gate was shut, as it was to prevent the loss of fertilizers and fish, circulation stopped, and the deeper water became stagnant and produced sulphuretted hydrogen. The great growth of phytoplankton caused the water to become too alkaline to be ideal for animal life, and for all these reasons the recovery of fish did not give an economic return.

Fertilization and the subsequent production of fish is, however, an industry hundreds of years old but it has been carried out more successfully in fresh water than in salt, and in the tropics than in temperate waters. C. F. Hickling's fascinating book on "Fish Culture" (1962) gives details of the practice in many lands.

The "stews" or ponds belonging to monasteries and great houses were at first only places to store live fish such as pike and perch for the winter but, by the 16th century, carp were being brought into England and fish culture was practised. The dung of sheep and cattle was recognized as a valuable

encourage this growth. When it has well started the depth of the water is increased to 12 cm and the milk-fish fry, about $1\frac{1}{2}$ cm long, are put in. They feed on the blue-green algae until they are 5 to 10 cm long when their food preference changes and they begin to prefer green algae. They are therefore transferred to larger ponds prepared in the same way except that the depth of water is originally made 10–15 cm and deepened to 50 cm or so when the fish are put in. In this greater depth green filamentous algae develop and on these the milk-fish now feed as well as on larger aquatic plants. In six to nine months the fish weigh 10–20 oz. Since the people are mostly poor and cannot afford to buy big fish they are sold off at this size and sometimes two crops a year are taken.

The culture varies in detail from one area to another. Where the milk-fish spawn twice a year four crops can be taken. Sometimes the small fry and fingerlings are kept separate from the older fish, sometimes they are reared together. Sometimes other fish with different food requirements are kept in the same pond to make full use of all available food. Prawns too, whose larvae come in with the sea water, are sometimes a valuable extra crop.

The other important fish in these brackish to salt-water pond cultures are the carp, mullet and *Tilapia*. This last is a genus with many species, some vegetarian, some carnivorous, and others omnivorous. Most of them live in fresh water and are widely used in fish culture in Africa and Israel; others are estuarine and can stand great changes in salinity.

Carp will tolerate only a small quantity of salt in the water (2·5%) and so can be used only in the least saline of the brackish-water ponds, but in Java the introduction of the carp *Puntius* into a milk-fish pond almost doubled the crop of milk-fish. It did so by eating the submerged non-algal vegetation, making this available to the milk-fish in its faeces and clearing the pond for the growth of algae.

Carp and mullet are grown in the brackish Hongkong fish ponds. The carp are freshwater fish able to stand only a low salinity, but the mullet are salt-water fish which can live in anything from fresh water to full sea water but which must return to the sea to breed. There is therefore a special fishery for the fry, as there is with milk-fish. The Hongkong ponds are heavily manured with soybean waste and night soil and, besides that, food such as silkworm pupae is added for the fish. The use of night soil is an insanitary practice and spreads parasites among the fish-eating population.

Although fish culture has been going on for centuries in the Far East, there is still much suitable land available for brackish-water ponds and, with the knowledge gained from the research work going on at the laboratories in Malacca and elsewhere, there could be a great increase in the fish-food for these over-populated and undernourished countries.

There are already many large-scale experiments in progress on the fertilization of open waters; the sewage of most coastal towns is discharged into the sea, adding annually tons of nutrient salts to the water. So far as can be seen these have little local effect: the phosphate content of the English Channel has decreased since the 1930's, in spite of the increase in population of the towns by its shores. Presumably the salts are too quickly dissipated over a wide area to have a measurable effect. But there are places where the discharge of sewage into a restricted area does have a distinct influence.

One such place is the Oslo fjord, receiving at its head the sewage from the town of Oslo. The fjord has always been an area of rich plankton and in recent years has had great "water blooms" which turned the water grayish-green or red according to the organism causing them. The coccolithophore *Coccolithus huxleyi* has risen in some years to numbers of many millions per litre and the dinoflagellate *Peridinium triquetra*, which occurs most years, to one or two millions per litre. These huge numbers are not all eaten by the zooplankton and sink to the bottom where they decay, using up the oxygen and killing off the fish population. To avoid this catastrophic effect a method has been developed and is being tried out, by which the phosphates are removed from the sewage by electrolytic action and can then be used as fertilizers on the land. Without the extra phosphate the phytoplankton would not form "blooms".

On the Mediterranean coasts of France there are numerous salt-water lagoons, shallow basins of salt water separated from the sea by a narrow bank of sand and kept from evaporating by canals from the sea. One of them is l'Étang de Thau which stretches for some 25 kilometres close to the sea and is connected with it by a canal at the town of Sète. Some interchange takes place through this canal, but as the tidal range is only some four inches the exchange is small. Into this lagoon go the waste waters of the town of Sète itself as well as those of several small towns on its banks. As a result there is a very rich phyto- and zooplankton, supporting a good fish population as well as rich beds of bottom-living shellfish. The most striking feature,

TABLE 3

The annual production of fish in various waters

Water	Main crop	Yield lb/acre/year
Unfertilized waters		
Alpine lakes	Salmonids	1·2–12·9
English and Welsh fresh waters	Salmonids	20·6
Ponds in Alabama	Bluegills	40–200
Brackish ponds, Philippines	Milk-fish	450–900
North Sea	Demersal fish	9
Barents Sea	Demersal fish	21
Iceland	Demersal fish	49
Peru	Anchoveta	400
Fertilized ponds		
Germany	Carp	260–400
Alabama	Bluegills	400–600
Israel	Carp	1890–2340
China	Carp	2800–6000
Belgian Congo	*Tilapia*	2500–10000
Animal crops on land for comparison		
Beef or mutton		200
Pork on cultivated land		500

however, is the oyster and mussel culture. All along the north shore, except in small gaps close to the towns, left for sanitary reasons, are stagings from which hang ropes supporting enormous numbers of mussels and oysters. The oysters, which are imported as young from Brittany, reach marketable size in a single year. The whole lagoon supports about 5000 people directly or indirectly by its fisheries. Twenty years ago there was little or no shellfish culture there, and the area taken up by this is now so great that there is friction between those who want to extend the oyster and mussel culture and those who want to catch fish or rake shellfish from the bottom.

A comparison of animal production in the sea and on land seems a suitable subject with which to conclude (see Table 3). Since from the land we eat mainly herbivores, and from the sea carnivores, the yield from land is bound to be much higher than from natural sea water, but the second may compare not unfavourably with the yield to man from wild animal populations on land.

With cultivated waters the yield depends largely on the length of the food chain. When this is at its shortest—with herbivores such as the grass carp, *Tilapia*, or milk-fish—very high yields indeed may be obtained, thousands of pounds weight per acre. To get these very high yields, however, much supplementary vegetable food has been added and the acreage which grew this should be taken into account. With carnivores even in fertilized ponds the yield drops to one or two hundred pounds an acre, and with secondary carnivores (fish which feed on other fish) it is less still.

On the other side of the picture, there is no comparison between the yield in plants useful to man from the two environments. A good crop of grass or corn amounts to several thousands of pounds weight per acre, but the return to man from plants in the sea is negligible.

What of the future? There is no doubt that more food can and must be got from the fertile sea to feed the hungry nations of the world. Inshore, in specialized localities there is room for development of fish and shellfish culture and much is already planned. For bottom-living fish we cannot at present hope to take bigger crops from near-by seas, but experiments on rearing plaice larvae through metamorphosis and then releasing them in the sea are promising. Greater yields in the future are likely to result from allowing the stocks on the ground to recover from the overfishing which has already taken place, and perhaps by controlling the competitive predators on the bottom so that the fish can get a larger share of the available food. Midwater fish, such as mackerel, tunny, and the herring and its relatives, are found in shoals and are often migratory, so that they are less easily exploited than bottom-living fish. Many of them, especially in the tropics, could be fished more intensively but in some cases (e.g. the anchoveta, the yellow-finned tunny and the North Sea herring) the fishery is already having too great an effect. Searching for the shoals in a wider area and the development of new types of ship and new types of gear seem promising lines for investigation. Above all the need is for research into all aspects of production in the sea.

References for Further Reading

J. S. Burlew (Ed.), 1953. *Algal Culture from Laboratory to Pilot Plant.* Carnegie Inst. Washington Publ. 600, 357 pp.

G. B. Deevey, 1960. "Relative effects of temperature and food on seasonal variations in length of marine copepods in some eastern American and western European waters." *Bull. Bingham oceanog. Coll.*, Vol. 17, pp. 54–86.

J. Fraser, 1962. *Nature Adrift.* G. T. Foulis & Co., London.

J. H. Fraser and J. Corlett (Eds.), 1962. "Contributions to Symposium on Zooplankton Production." *R. P.-v. Réun., Cons. perm. int. Explor. Mer*, Vol. 153, 228 pp.

M. Graham, 1943. *The Fish Gate.* Faber and Faber, London.

A. C. Hardy, 1956. *The Open Sea. The world of plankton.* Collins, London.

A. C. Hardy, 1959. *The Open Sea. Fish and fisheries.* Collins, London.

A. C. Hardy and R. Bainbridge, 1954. "Experimental observations on the vertical migrations of plankton animals." *J. mar. biol. Ass. U.K.*, Vol. 33, pp. 409–448.

A. C. Hardy and E. R. Gunther, 1935. "The plankton of the South Georgia whaling grounds and adjacent waters 1926–27." *Discovery Rep.*, Vol. 11, pp. 1–456.

H. W. Harvey, 1955. *The Chemistry and Fertility of Sea Waters.* Cambridge University Press.

C. F. Hickling, 1962. *Fish Culture.* Faber and Faber, London.

International Council for the Exploration of the Sea, 1958. "Contributions to Symposium on Measurements of Primary Production in the Sea." *R. P.-v. Réun., Cons. perm. int. Explor. Mer*, Vol. 144, 158 pp.

N. G. Jerlov, 1951. "Optical studies of ocean waters." Rep. Swedish deep-sea Exped., 1947–48, Vol. III, pp. 1–59.

S. M. Marshall and A. P. Orr, 1955. *The Biology of a Marine Copepod, Calanus finmarchicus* (Gunnerus). Oliver and Boyd, Edinburgh.

J. H. Orton, 1937. *Oyster Biology and Oyster Culture.* Buckland Lectures, Arnold & Co., London.

M. Parke, also with I. Manton, B. Clarke and D. G. Rayns, 1949, 1955, 1956, 1958, 1959, 1962, 1964. "Studies on marine flagellates." I–VII. *J. mar. biol. Ass. U.K.*, Volumes 28, 34, 35, 37, 38, 42, 44.

L. Provasoli, J. J. A. McLaughlin and M. R. Droop, 1957. "The development of artificial media for marine algae." *Arch. Mikrobiol.*, Vol. 25, pp. 392–428.

L. Provasoli, K. Shiraishi and J. R. Lance, 1959. "Nutritional idiosyncrasies of *Artemia* and *Tigriopus* in monaxenic culture." *Ann. New York Acad. Sci.*, Vol. 77, pp. 250–261.

J. E. G. Raymont, 1963. *Plankton and Productivity in the Oceans.* Pergamon Press.

F. S. Russell, 1925–1934. "The vertical distribution of marine macroplankton I–XII." *J. mar. biol. Ass. U.K.*, Volumes 13–17, and 19.

W. H. Schuster, 1952. "Fish culture in brackish-water ponds of Java." *Indo-Pacific Fish. Coun., Spec. Publ.* 1, 143 pp.

E. Steemann Nielsen, 1952. "Use of radioactive carbon (C^{14}) for measuring organic production in the sea." *J. Cons. perm. int. Explor. Mer*, Vol. 18, pp. 117–140.

H. U. Sverdrup, M. W. Johnson and R. H. Fleming, 1942. *The Oceans.* Prentice-Hall, New York.

G. Thorson, 1958. "Parallel level-bottom communities, their temperature adaptation and their 'balance' between predators and food animals." In *Perspectives in Marine Biology.* Ed. Buzzati-Traverso. University of California Press.

R. S. Wimpenny, 1953. *The Plaice.* Buckland Lectures, 1949. Arnold & Co. (Fishing News (Books) Ltd.), London.

C. M. Yonge, 1960. *Oysters.* Collins, London.

INDEX

Figures in bold type denote illustrations